How to Save the Internet

Also by Nick Clegg

Politics
How to Stop Brexit

NICK CLEGG

How to Save
the Internet

*The Threat to Global Connection in
the Age of AI and Political Conflict*

THE BODLEY HEAD
LONDON

1 3 5 7 9 10 8 6 4 2

The Bodley Head, an imprint of Vintage, is part of the
Penguin Random House group of companies

Vintage, Penguin Random House UK, One Embassy Gardens,
8 Viaduct Gardens, London SW11 7BW

penguin.co.uk/vintage
global.penguinrandomhouse.com

First published by The Bodley Head in 2025

Typeset in 11.5/15pt Sabon LT Std by Jouve (UK), Milton Keynes
Printed and bound in Great Britain by Clays Ltd, Elcograf S.p.A.

The authorised representative in the EEA is Penguin Random House Ireland, Morrison
Chambers, 32 Nassau Street, Dublin D02 YH68

A CIP catalogue record for this book is available from the British Library

HB ISBN 9781847928597
TPB ISBN 9781847928603

Penguin Random House is committed to a sustainable future
for our business, our readers and our planet. This book is made
from Forest Stewardship Council® certified paper.

To Miriam, Antonio, Alberto and Miguel – the best fellow travellers anyone could wish for

Contents

PROLOGUE

You are living through one of the most rapid periods of techno-logical change in centuries. It is barely three decades since the pinging and popping of dial-up modems brought the internet into our homes via bulky desktop computers; now it is embed-ded in every aspect of our lives. The iPhone wasn't released until 2007; now many of us struggle to imagine getting by without a smartphone. Facebook launched in 2004; now more than twice as many people use it as drive cars. Somewhere along the line, Google became a verb. A hundred million people were using ChatGPT within two months of its launch. More than a hun-dred billion messages are sent every day on WhatsApp alone. Even the most technologically inept baby boomer has more information, more choice, more people – and more power – at their fingertips than their grandparents could have dreamed of.

The open, borderless and largely free internet has become an integral part not only of our daily lives but also of how our societies function. Processing data at scale has revolutionised the way businesses operate and public services are delivered. People generating and sharing content at previously unheard-of scale and speed has transformed the public sphere. From the Arab Spring to Donald Trump's breakthrough election in 2016, social media has been seen as a driving force behind some of the big political upheavals of recent times. At the same time, the ability to connect with each other online kept the wheels of society turning during the Covid-19 pandemic in a way that would have been unthinkable just a few years before. To quote an internet meme of Will Ferrell in *Anchorman*: 'That escalated quickly.'

For as long as there has been digital technology, there have been people who have cautioned us against it, but in recent years many societies – especially in the West – have become increasingly anxious about the size and power of tech companies, how our data is held, used and monetised, and the impact internet technologies have on everything from our privacy and mental health to the exercising of free speech and the functioning of democracy. In many quarters it has become something like accepted wisdom that Big Tech is exacerbating or even at the root of many of society's ills.

This backlash, or 'techlash', raises difficult questions. Would we choose to turn off certain technologies if we could? Would we deny ourselves the next innovation? Would we be better off without the freedoms of the open internet? Some would say yes. Most of us, I suspect, experience a dilemma: fundamentally we recognise and welcome the benefits that technology brings, but we baulk at the fact that they put power in the hands of people and entities we don't trust and don't feel are accountable to us. As I hope to show in this book, how we attempt to resolve this dilemma is a question of immense importance that will reshape our relationship with the internet.

As I also hope to show, it is no coincidence that this techlash has come to a head during a period when the international order is pulling apart. Since the 2008 financial crash, the politics of economic nationalism and geopolitical rivalry have been turbocharged. Trump's 'America First' reliance on tariffs and trade wars has up-ended the global economic order and antagonised allies and adversaries alike. Both democracies and autocracies have become increasingly insular, seeking to exercise greater sovereignty over their economies, their borders and the technologies used by their citizens. The result is a new wave of laws and regulations as nation states seek to reassert control over the internet.

Donald Trump's return as President of the United States promises to accelerate this trend. A significant difference from his first term is that, this time around, Silicon Valley tech leaders have sought to align themselves with Trump and his forceful America First philosophy. This is at least in part based on the belief that his administration can exert pressure on foreign governments to refrain from onerously regulating American companies. Shortly after I announced my resignation from Meta in January 2025, the company caused controversy by shifting its stance on content moderation, including loosening some of its restrictions on what content is allowed on its platform, and replacing its independent fact-checking programme in the US with a crowdsourced X-style 'community notes' approach. Mark Zuckerberg made the politics of the shift clear when he announced the changes, saying:

> We're going to work with President Trump to push back on governments around the world going after American companies and pushing to censor more. The US has the strongest constitutional protections for free expression in the world. Europe has an ever-increasing number of laws institutionalising censorship and making it difficult to build anything innovative there. Latin American countries have secret courts that can order companies to quietly take things down. China has censored our apps from even working in the country. The only way we can push back on this global trend is with the support of the US government.

While it's too early to tell how this assertion of techno-unilateralism in the United States will play out – uniting Silicon Valley and Washington DC against foreign laws and norms – two things seem most likely: first, that given America's extraordinary economic and technological weight the US government will succeed on some key issues in imposing its will on others; second, that it is equally certain that over time there will be a reaction from sovereign countries elsewhere, if for no other reason than that it will be politically impossible in places as varied as India, Brazil, Indonesia and France simply to meekly do what they're told by Uncle Sam.

So the new rules of the internet are being written at a time when the international order is unravelling, with superpowers flexing their muscles on the world stage and nations of all shapes and sizes determined to assert greater sovereign control over their economies and their ways of life. Had the techlash occurred in the political climate of a generation ago, it is conceivable that this spate of policymaking might have been approached differently. The internet is by its very nature global. Every nation state is grappling with versions of the same problems and concerns. In the era of globalisation, when international institutions like the United Nations or the World Trade Organization were at the height of their powers and even many autocratic states were seeking to open up their economies, it is conceivable that international treaties or other global mechanisms could have brought a level of cohesion to the way the rules of the internet were written and applied. Instead, nation states are ploughing their own furrows, creating a complicated patchwork of local laws with little or no attempt to coordinate across jurisdictions. The net result is that the global internet is splintering into national and regional silos, with potentially huge knock-on effects for the global economy, the trajectory of technological innovation, and ultimately the freedoms and opportunities that will shape billions of lives.

In fact, the global internet in its truest sense no longer exists. China, Russia and others have built digital walls at their borders, effectively segregating their online worlds and creating an alternative internet model. For those of us living in western democracies, the only version of the internet we have ever known is one in which a piece of information can reliably be sent from one digital address to another without any government deciding how – or whether – that happens. In this alternative model, nation states exercise top-down control over information and citizens are cut off from the world beyond their borders. The temptation for other nations – democracies and autocracies alike – to follow suit has only increased with time.

We have arrived at a crossroads – and at a crucial moment. Just as the techlash and deglobalisation are changing the character of the internet from the outside, powerful advances in generative AI are changing it from within. If we continue along our current path, we will soon find ourselves in a world where the internet has changed beyond recognition. As countries develop different versions of the internet, segregated by national or regional borders, divergent AI technologies risk deepening and entrenching the divisions between and within our societies. As fears about the perils of technologies like social media and AI dominate public discourse, and with the political world reacting by seeking to assert national control over the once borderless internet, we risk losing what has made it so liberating. As I set out in the second half of the book, this will have hugely damaging effects not only on global trade and economies everywhere, but also on medical and scientific research, cross-border policing, academia, journalism and more. Ultimately, the world will be less innovative, less open and less free.

But there is an alternative path, one that could lead us away from this divided future by salvaging and protecting the best of the global internet and ensuring that the benefits of AI are spread to the largest possible number of people. At its heart, the internet is a revolutionary experiment in openness: from the way its early pioneers hardwired the open flow of information into its design, to the way it has opened up access to the public sphere to billions of people, creating a messy but ultimately empowering marketplace of ideas, creativity and commerce. Openness, I believe, is also the key to saving it.

This will require the democratic world to make common cause around our technological future. In reality, this means the United States needs to recognise that its own best interest lies in pooling some of its technological advantage with a coalition of nations that share its democratic values – not least India and the European Union. In the final chapter, I propose a US-led global deal to keep the flow of data between these techno-democracies open and to share access to vital AI infrastructure, establish global standards for the technology's responsible development, and enable cooperation between allies on sensitive areas like security and intelligence. Such a deal may seem counter-intuitive at a time when President Trump's America is escalating its trade war with China, imposing or threatening tariffs on long-term allies like Canada, Mexico and the European Union, and when governments in India, Europe and elsewhere are imposing ever greater sovereignty over the internet within their national borders. But I believe there are strong political incentives for all sides that make it plausible.

For any of this to happen, we first need to forge a new relationship between tech and society. To save the open internet, I will argue that Big Tech itself is going to have to become a lot more open. A small number of private companies have accrued immense economic and political power in recent years, and this power needs to be wielded transparently and held to account. In the penultimate chapter, I will propose some concrete steps that Big Tech companies can take to open up their platforms to much greater scrutiny, devolve decision-making power over content that is shared online, bring their users into their policymaking processes and, crucially, give them far greater control over their online experience – including how their data is used.

It may seem unrealistic to expect companies that have spent years amassing power to voluntarily start giving it away. But it is in their long-term self-interest to do so. If the masters of Silicon Valley refuse to open up, the choice will be taken out of their hands. The political world has many ways to assert sovereignty over private enterprises – not least by breaking them up or even one day nationalising the vast AI infrastructure that they have spent billions upon billions of dollars building. These sorts of drastic government interventions would be bad, for a number of reasons I will set out later in this book, but they could prove hard to resist for politicians eager to assert control over Big Tech. The best hope for avoiding them is if Big Tech lets the sunlight in itself. We need a radical level of transparency and accountability, which in turn will provide the basis for well-informed, multilateral rule-making underpinning the open internet.

Fundamentally, this is a book about power. Future historians will decide where the digital revolution of the late twentieth and early twenty-first centuries ranks alongside the likes of the Industrial Revolution, the Reformation and the other great societal upheavals of history. But it will be up there with them. And like these other great historic disruptions, the digital revolution has disrupted power – shifting it away from many of the traditional gatekeepers and institutions of power, to new people and new institutions. In doing so it has created a paradox that goes to the heart of many of today's biggest debates about technology: on the one hand, digital technologies, from the internet and social media to generative AI, have greatly democratised power, and promise to continue to do so, loosening the control of political and media gatekeepers, leapfrogging geographic boundaries, and giving billions of people access to information, economic and social opportunities and the ability to express themselves in ways that were once the preserve of a privileged elite; on the other, they have centralised economic and political power in the hands of a small number of vast internet platforms.

This power paradox raises profound questions. How do we curb the power of the platforms without also curbing the power of the people? What does Silicon Valley need to do to ensure it maintains society's permission to innovate? Can nation states ever truly control the internet within their borders; and, if so, at what cost? What level of risk are we prepared to accept in return for the extraordinary benefits that are available to us from AI, and how do we go about creating those rules in a way that ensures everyone buys into them?

These are not questions that any single book or individual can or should resolve alone. But we urgently need to come together to negotiate answers to these questions. My hope is that this book might provide some insight into how we could do that, and why it is so essential that we do.

But first, I should tackle a more immediate question: why should anyone listen to me?

INTRODUCTION

The Corridors and Campuses of Power

I'm not a creature of Silicon Valley. I didn't study computer science. I've never written a line of code. I haven't spent my career – the bulk of it anyway – immersed in the processes, debates and science of technological progress. I came to Meta, Facebook as it then was, in 2018 as an émigré from the world of British and European politics and stayed until 2025. In my twenties I worked in the European Commission as a trade negotiator before embarking on a career in elected politics in my thirties. I was elected first as a Member of the European Parliament, then as a Member of Parliament in the United Kingdom. In late 2007, I became Leader of the Liberal Democrats, Britain's third party. Just over two years later I led the Liberal Democrats into an election that resulted in a hung parliament (where no party has an overall majority) and ultimately became the deputy prime minister in the UK's first coalition government since the Second World War. It wasn't until three years after that government ended that I first set foot in Silicon Valley. I wasn't really sure what to expect.

'Silicon Valley' sounds like a made-up place – a nickname or shorthand rather than an actual geographic location. In reality, it describes a long stretch of suburban California that starts where urban San Francisco ends on the west side of San

Francisco Bay, then runs south along the route of Interstate 280, bordered on one side by the bay and the mountains of the Diablo Range, and on the other by the Santa Cruz mountains (hence the valley in the name). Further north is wine country. Further east is desert.

It's in the towns that make up this sun-kissed stretch of suburbia – Palo Alto, Menlo Park, Mountain View, Cupertino and many more – that generations of tech start-ups have burgeoned in the hope of becoming the next big thing. And many of them did. Silicon Valley today is a beacon attracting software engineers and venture capitalists alike, all hoping to strike it rich with the next billion-dollar company. The streets of Palo Alto are full of young, smart, wealthy people from every corner of the US and the world, hanging out, wearing hoodies, carrying water bottles with slices of cucumber inside and driving eerily quiet Teslas.

As an Englishman, I never cease to be amazed by the sheer scale of America. It's striking, when you first set foot in this part of Northern California, how remote it feels from the power centres of the east coast. It's not just the nearly 3,000 miles that separate you from Washington DC and New York; the three-hour time difference means you feel oddly out of sync. The day's news agenda is in full flight on the east coast before the west is awake. The biggest European stories broke the night before. Then, in the evening in California, it feels like the rest of the world is asleep. Californians watch the New Year being greeted with the Times Square Ball Drop, then enjoy three more hours of the previous year – or watch it three hours late as a re-run. It's like you're far away from everywhere.

It's no wonder that people who want to strike out on their own, at a safe distance from the prying eyes of the men in suits, are attracted to this place. It's a natural home for idealists who want to stick it to the man. This part of the world has been a haven for adventurers and anti-establishment figures for generations. The world's tech centre is in the heart of gold rush country, and not far from where, a century or so after the gold rush, the hippie movement started in San Francisco's Haight-Ashbury district.

This might explain why Silicon Valley technologists are prone to talking about their work in visionary terms, perhaps best captured in the writer and cyberspace evangelist John Perry Barlow's end-of-the-millennium manifesto *A Declaration of the Independence of Cyberspace*. As he somewhat loftily put it: 'Governments of the Industrial World, you weary giants of flesh and steel, I come from Cyberspace, the new home of Mind. On behalf of the future, I ask you of the past to leave us alone. You are not welcome among us. You have no sovereignty where we gather.' There's something about the character of the place that lends itself to this libertarian utopianism. And if you're looking for a more direct connection between technologists and hippies, Barlow could occasionally, when not extolling the internet's freedom from authority, be found writing lyrics for the Grateful Dead.

There's something intoxicating about Silicon Valley. This is a place where people say yes. It's a magnet for smart people buzzing with creativity. There are no limits to ambition. Every obstacle is an opportunity. Every beautiful day could be the day something amazing happens. Anyone could have an idea that could change the world – and get filthy rich. If you build it, they will come. And the things that get built here – by these idealistic people in this dizzying environment – really do change the world. This is the home of modern computing, where the microprocessor and the personal computer were created, where the World Wide Web took off, as well as the first web browsers and search engines that helped us make sense of it, right through to social media, smartphones, cloud services and now generative AI.

But it is also an industrial-scale breeding ground for hubris, of a largely macho kind too (it is still a place where, on the whole, confident men with big egos rule the roost and smart, capable women have to fight for their place every step of the way). The isolation that makes it the perfect petri dish for new ideas is also what distances it from the lives of ordinary people. The wealth of the place distances it from their struggles. Its idealism distances it from the messy realities of their lives.

One of the biggest culture shocks I experienced in Silicon Valley had nothing to do with differences between the UK and the US, per se. It was because the worlds I had moved in ever since I first left home for university in the mid 1980s – studying social anthropology at Cambridge, and throughout my brief career in journalism and my significantly longer one in politics – were ones where the primary way of succeeding was through the power of words. Politics, at its heart, is a competition between different stories about how things should be. That's what liberalism, socialism, conservatism and fascism ultimately are: stories that give us different ways of looking at the world, different diagnoses of our problems, and different solutions too. You win elections if more people sympathise with your telling of events and your vision of the future than with those of your opponents.

But Silicon Valley is full of people who see the world a different way: engineers. Theirs is a world of facts and process. The engineer's mindset is to identify a problem and fix it, then move on to the next problem. They operate in a maze of acronyms, as if language itself is a time-wasting exercise that needs to be boiled down to its most abbreviated form. On arrival, I was immediately plunged into a world of XFNs, STOs and FOAs, where every roadmap has a playbook, every community is part of an ecosystem, and where every subject is either a tl;dr or a deep dive (but never, seemingly, the right length). And everything – everything – has to be quantified. In an early meeting, I remember being asked by one of the company's most senior engineers what the percentage likelihood was that government X would pass a law about Y. I laughed. I thought it was a playful joke. The idea that political processes could be boiled down to a sort of faux-science seemed silly to me. He didn't laugh. So I said something like, 'Oh, well, 23.67 per cent.' He nodded earnestly. I have since come to realise that in order to persuade people in Silicon Valley, compelling stories are useless without data points for every argument and probabilities for every outcome. And I must admit, it has rubbed off on me. Approaching problems in a systematic way, however faux the science might be, helps to order your thought process and guide you towards clear decisions.

But it is also reductive. It commodifies everything, removes nuance and instinct and all the intangible aspects of human nature. When I talked to engineers in my early days at the company about data, they talked about it the way a carpenter would talk about wood – a material to be used to make amazing things, not people's precious personal information. That mindset has changed in the years since, in part because of a wider cultural shift in people's expectations of privacy and data protection, and in part because change has been forced upon the company by regulators. But the primacy of logic and reason over emotion and gut instinct gives a clue, perhaps, as to why Silicon Valley can often seem a little tone deaf to the concerns of people beyond the walls of its primary-coloured campuses.

Perhaps inevitably, now that we are a generation or more deep in the internet revolution, the early utopianism of Silicon Valley has turned from sounding visionary and inspiring to occasionally appearing naive and self-serving. Google's infamous 'Don't Be Evil' motto was never meant to be ironic. But it's hard to sense the winds of public opinion changing direction when you're sunning yourself in Santa Cruz.

In a way, Silicon Valley has become its own worst enemy. When the financial crash and its economic and political aftershocks were reshaping the agenda, companies that were built to sell sunlit uplands were painfully late to the change in mood. The lofty idealism started sounding hollow. The eye-watering amounts of money being generated made tech companies obvious targets as governments wrestled with growing deficits and people faced rising costs and cuts to their public services. Tech leaders made missteps, failing to distinguish clearly between legitimate concerns and trivial ones. Convinced of their own anti-elitist self-image, they soon cemented themselves in the public imagination as a new class of out-of-touch elite. Nerdy heroes transformed into capitalist villains.

Suddenly the Silicon Valley tech bro is the archetypal villain of movies, from blockbusters like *Jurassic World* and *Glass Onion* to animated movies like *The Mitchells vs The Machines* and *Ron's Gone Wrong*. A *Superman* reboot sees Lex Luthor reimagined as a tech bro, played by the same actor – Jesse Eisenberg – who shot to fame playing the young Mark Zuckerberg. Tech bros are the villains du jour, just as mad scientists with German accents, Russian hard men, heartless Wall Street bankers and shady Middle Eastern terrorists were before them.

Still to this day, tech leaders don't always help themselves. During a virtual session of the Seoul AI summit in 2024 – the follow-up to the 2023 UK-led Bletchley Park summit in Milton Keynes – which brought together leaders of the G7 nations, South Korea and others with business leaders from the big AI companies to discuss the responsible development of the technology, Elon Musk appeared live from his private jet. Hunched over his laptop, front-lit against a dark background, he appeared as if in a hostage video shot on the Death Star, before reprimanding everyone about how AI needs to be 'maximally truthful' and not focused on the 'fashion for social justice'. An instant later his link went down, leaving world leaders and diplomats staring nonplussed at their Zoom screens. Eventually he came back online, declared 'honesty is the best policy', and promptly disappeared again. Another tech leader appeared to be eating his breakfast throughout the summit.

As a recovering politician, I know a thing or two about being caricatured as a villain. I was in government for five years – from 2010 to 2015 – as deputy prime minister in a Liberal Democrat–Conservative coalition. For all the achievements (and there were many) of the coalition government overall and of the Liberal Democrats within it, they came at a high political price for the party and for me personally. I had a brief moment in the sun during the 2010 election campaign when, in the wake of introducing myself to the British public in the UK's first ever televised leaders' debate, I was fleetingly the most popular politician in the country. But I was soon knocked off my perch, even becoming something of a hate figure on both sides of the aisle, accused by the right of holding the Conservatives back from enacting a true-blue agenda, and by the left of betraying my principles and enabling the hated Tories. I've been protested against, satirised, vilified by both the right-wing and the left-wing press, even hanged in effigy. So one result of my time as a politician is that I know intimately what it means to lose control of the narrative.

What I also learned from being in government is something of the way that technology is viewed by the political class: how technological advances resuscitate age-old political dilemmas like the tension between privacy and security, or individual liberty and top-down state control. Above all, it revealed to me a significant disconnect between the sometimes politically naive technologists building these remarkable innovations and the often technologically illiterate politicians charged with regulating them.

The government wants to spy on you

I remember encountering, relatively early on in my time as deputy prime minister, a consistent temptation within the government to seek to curtail technology and/or to use it for surveillance of people in the name of national security. As I discovered, the British security apparatus is adept at taking wet-behind-the-ears ministers and exposing them to chilling presentations of numerous new threats in order to instil the requisite alarm to justify whatever draconian policies are contained on the next slide. In opposition during the Blair–Brown years, I'd campaigned against Labour Home Secretaries – the cabinet ministers in charge of what Americans refer to as Homeland Security – trying to introduce illiberal measures like the introduction of compulsory ID cards and up to ninety days' detention without charge for terror suspects.

Shortly after I entered government, the Home Office argued that the move from itemised billing of phone calls to flat-rate monthly contracts and unmetered internet use was a national security concern. This is because, historically, telecoms companies collected and stored every number you called in order to bill you, based on time and distance. This meant the 'communications data' (whom you called, when, from where, and for how long – though not the actual content of the message) was also, coincidentally, available (with the appropriate permissions) to the police and the intelligence agencies. With the spread of mobile phones in the early 2000s, companies had increasingly done away with itemisation and as a result no longer had a business case to collect the data. The mobile internet was accelerating this trend – lots more data, but very little of it available for law enforcement or surveillance. The Home Office wanted to pass legislation – a Communications Data Bill – to shore up the previous model, requiring telecoms companies to effectively act on the state's behalf in collecting information about their customers' communications for which they themselves had no use. This would also be extended to text messages, emails and web browsing, including social media activity. The press predictably and accurately enough dubbed the bill the 'Snoopers' Charter'.

Every proposed piece of legislation that hadn't been included in the original coalition agreement – the programme for government hammered out by the Conservatives and Liberal Democrats prior to forming the government – had to be agreed by both parties. The Snoopers' Charter wasn't part of the coalition agreement, which ultimately meant that it couldn't proceed if I couldn't support it. The Conservatives and the security services were strongly in favour and applied considerable pressure on me and my party's ministers. We gave only partial consent, insisting the legislation should be published in draft rather than in full, and sent to a special committee of both Houses (the Commons and the Lords) to be scrutinised, including by liberals who were wary of government overreach on issues of civil liberties. The committee published a critical report with a number of recommended changes to the bill, which became the subject of heated debates between the coalition parties. Ultimately, I insisted the bill couldn't proceed unless it was rewritten substantially. When it became clear this wasn't going to be done in a way that we felt was palatable, I eventually blocked the bill in April 2013.

A few weeks later, in June 2013, the *Guardian* and the *New York Times* began publishing a large number of top-secret documents leaked by US National Security Agency (NSA) contractor Edward Snowden, revealing an array of sophisticated surveillance programmes. The reports of mass surveillance were jaw-dropping. The NSA had been collecting data on millions of innocent people's phone and internet activity without their knowledge or consent. Buried in Snowden's anarchic data dump was evidence of the existence of top-secret surveillance programmes like PRISM, which allowed the NSA to collect data from tech companies including Google, Facebook and Apple. And the UK was implicated too. The leaks revealed that the UK's intelligence agency Government Communications Headquarters (GCHQ) had been working closely with the NSA to collect and analyse data on people's phone and internet activity. One of the programmes exposed by Snowden was called Tempora, which allowed GCHQ to tap into transatlantic fibre-optic cables and collect vast amounts of data on internet traffic, including emails, web browsing history and other online activity.

One thing that was painfully obvious was that the Home Office and the UK's intelligence service at GCHQ had withheld significant information from me and other democratically elected ministers. We learned far more about their work from reading the *Guardian* than we ever did from them. It severely deepened the mistrust between us and the Conservatives when relations were already at a low ebb after I had nixed the Snoopers' Charter. We pushed them to agree to a review of surveillance powers, which they resisted. They eventually relented in July 2014, when the Home Office asked David Anderson QC – a respected barrister who served as the Independent Reviewer of Terrorism Legislation – to lead a review. He produced a 300-page report the following year which formed the basis of a new, comprehensive Investigatory Powers Act, passed in 2016.

These and other experiences in government left me with a bad taste in my mouth. But they also piqued my interest. Throughout my time in government I would meet representatives from companies like Google or Microsoft to understand how they grappled with government data collection demands. To my surprise, I often found myself more sympathetic to the technologists than the government. These were global companies with a global outlook, trying to navigate a new and increasingly fraught area of politics and strike a balance between liberty and security, and seemed to be doing so with greater appreciation of the trade-offs than the Home Office itself.

In my life after government, I found myself increasingly curious about the interaction of tech and politics. Through Open Reason, a think tank I established after leaving government, I dipped my toes into issues around social media, artificial intelligence and tech regulation. I talked to entrepreneurs and start-ups about the challenges they faced, and those they hoped to help solve. I began to explore the idea of creating a forum where up-and-coming politicians could meet and engage with innovators and businesses in the tech world. I wanted to find a way to bridge the divide between what seemed to me to be a world of technologists who didn't speak the language of politics, and the world of politics that didn't remotely understand the technology it was supposed to be regulating.

It was while I was tiptoeing into these debates that a series of events began to unfold that would come to be foundational moments in the development of the techlash – and Facebook was at the centre of them. In the aftermath of Donald Trump's victory in the 2016 US Presidential election, it became clear that entities based in Russia had attempted to use social media platforms like Facebook to sow disinformation, including spreading fake news and running ads aimed at American voters using bogus accounts linked to Russia's notorious Internet Research Agency bot farm. This was followed by a second scandal, when it was revealed that an academic named Aleksandr Kogan had sold data related to millions of Facebook users – data that he had access to strictly for academic and not commercial purposes – to a political advisory firm called Cambridge Analytica, which then claimed to have used that data to help Mr Trump's campaign target US voters. (Widespread claims that the firm also helped the Vote Leave campaign to target British voters during the 2016 Brexit referendum were later debunked by the UK's Information Commissioner's Office.) Around the same time, Facebook was also alleged to have been used to spread hate speech and incite violence against the Rohingya people in Myanmar during a genocidal campaign by the military.

Suddenly Big Tech – and Facebook in particular – was in the firing line in a way it hadn't been during my years in office. Big problems had been exposed. Silicon Valley had lost the benefit of the doubt. Not only was it under scrutiny as never before, but all good faith had been lost.

The unwinding of globalisation

I'm very much a child of globalisation. I grew up in leafy middle England, but I've always considered myself European. My mother is Dutch and my father has Russian heritage. As a young man I studied at the College of Europe in Bruges, where I met Miriam, my wife, a proud Spaniard. We raised our three boys in a bilingual house. In my time as a trade negotiator in the European Commission and later as an MEP representing a region of one country in a vast parliament with twenty-four other Member States, I was on the very front lines of globalisation.

The coalition government came to power in the wake of two seismic political events. The first was the financial crash. I was less than a year into leading the Liberal Democrats when the 2008 crash happened – the collapse of Lehman Brothers occurred on the same weekend as my first autumn party conference as leader. The banking crisis was a cataclysmic event, causing hardship and heartache for millions in the years that followed. Developed economies are still dealing with its aftershocks today. It has been a wrecking ball for the UK's public finances, leaving a long tail of unsavoury economic repercussions for the coalition, and every government since, to deal with.

The second event, probably unknown to readers outside the UK, but also of seismic importance within its shores, was the 2009 MPs' expenses scandal, which laid bare the misuse of public funds – some egregious, some trivial – by dozens of British politicians of all parties, which eroded further what little trust the British public had in their political class.

I don't claim to have seen the backlash against globalisation and the rise of anti-establishment populism coming. But I did have a front-row seat. In many ways, my political rise during the 2010 general election was fuelled by popular anger at Britain's two old establishment parties. The politics I preached was one of change and political reform, for which there was a ready appetite among voters fed up with the status quo. And when I found myself in government, I saw close-up how our governing partners, the British Conservative Party, became ever more entranced by populist nationalism.

The David Cameron with whom I negotiated the coalition was a self-described 'liberal conservative', campaigning on the National Health Service and his environmental credentials (even being photographed hugging huskies in the Arctic to underline the point). But soon the populist headbangers on the right of his party were exerting an ever greater grip over its politics. The UK Independence Party, an outright nationalist party, rose in popularity and influence, and chewed away at the Conservatives' base of support. As it did, right-wing Conservative MPs and their cheerleaders in the media and the party's grassroots became ever more vocal. To keep them on side, Cameron and his Chancellor, George Osborne, threw them ever more red meat – from proposing ever harsher cuts to welfare spending (which I spent considerable political capital blocking), to giving tax breaks to married couples (which I agreed to in exchange for the introduction of free school meals for 4–7-year-olds) – culminating in his promise to hold a referendum on Britain's membership of the European Union if he won a majority of his own in the 2015 general election, which he duly did. Despite pandering to the Brexiteers in their party, Cameron and Osborne campaigned for Britain to remain in the EU, alongside the Liberal Democrats and most prominent Labour politicians. But the battle was lost.

Brexit was heartbreaking for me. My identity as a European is a fact of my family history, just as Britain's place in Europe is a fact of geography. I've devoted much of my career to the idea of European solidarity and the institutions of European cooperation. As a politician, I have advocated passionately that we are stronger together than apart, better able to solve shared challenges when we work collectively than in isolation. I believe this not just for Europe but for the world, which makes me what some might describe as a globalist – though that word is almost exclusively used as an insult these days. But I mean it in the sense that I believe in international cooperation. I believe that co-dependency and mutual shared interest between nations is the best way to avoid conflict, and that pooling a small portion of national sovereignty in order to act collectively is the only way we will rise to truly global challenges like climate change and cross-border crime. It is also the best way, in my view, to manage a borderless entity like the internet.

Glutton for punishment

By now it may be clear that I am a serial holder of unpopular positions. I was the leader of the third party in a country that had been led by one of two larger parties – Labour and

Conservative – for about a hundred years. I was the deputy prime minister in a coalition government in a country that hadn't had a coalition since the Second World War. I'm a European from a country that voted to leave Europe. I'm a liberal centrist in an era of populism on both the right and the left. I'm a staunch internationalist at a time of deglobalisation and resurgent nationalism. And, being the glutton for punishment that I evidently am, I then worked for Meta at a time when social media was being blamed for many of the world's ills. But, however unfashionable an opinion it may be, I believe that progress happens when people are empowered, not when politicians, the media, elites or anyone else decree it, and in my view that, fundamentally, is what the internet and social media are all about.

Even so, while I had my curiosity piqued by my experiences in government and afterwards, I am probably not the sort of person you would expect to find in the gleaming campuses of Silicon Valley. I spent my career in a suit and tie, not a hoodie and flip-flops. My first instinct when confronted with the hundreds of brightly coloured sloganeering posters that adorn the walls of every meeting room and corridor of Facebook's extraordinary MPK campus (another acronym, this one a somewhat unnecessary shortening of Menlo Park), was to do something very English: apply gentle mockery. In one of the first meetings I held with one of the teams I had just taken charge of, a poster on the wall declared the ubiquitous Silicon Valley mantra: 'Bring Your Authentic Self to Work'. To try to break the ice, I said, 'Please don't bring your authentic self to work. You wouldn't like my authentic self if I brought it to work. So just bring your inauthentic self to work from nine to five and then you can go home and be yourself and we'll get on perfectly well.' Stony silence. One of the team came up to me rather coyly afterwards to say that the message had been quite disconcerting for them. I knew then that I wasn't in Kansas any more.

The fact that I was approached for the role at all is a recognition that what was then Facebook is not simply a technology company. Its sheer scale and the ubiquity of its products, coupled with it finding itself in the crosshairs of public debate, make it an inherently political institution too – regardless of whether it wants to be or not (it doesn't). The techlash has meant its ability to operate is not simply a question of its inventiveness, technical know-how and financial muscle, but of its role in society. Put simply, the techlash has thrown into doubt whether Meta, Google, Apple, Amazon and other tech companies have society's blessing to do what they do. To ensure that these companies can continue to innovate in the years and decades ahead, they have realised that they need to be in a dialogue with the world of politics, not a shouting match.

It was in the aftermath of Facebook's 2016–2018 controversies that I had a series of conversations with leaders at the company. My friend and predecessor as MP for Sheffield Hallam, Richard Allan, introduced me to Elliot Schrage, Facebook's then (and soon-to-depart) Vice President for Policy and Communications. Before long I was approached by its Chief Operating Officer, Sheryl Sandberg, and flown out to California to meet her and Mark Zuckerberg.

It was obvious to me from our first encounters that Mark and Sheryl were a formidable duo. It is commonly understood that Mark Zuckerberg is a visionary innovator, but the two things that struck me most about him are his endless curiosity and his indefatigable competitiveness. Lots of people who achieve great personal success, especially at a young age, can tend to get stuck in their ways and rest on their laurels. Mark, on the other hand, has the humility, drive and appetite to keep learning. If he thinks he doesn't know something, he will grill the people who do and devour every bit of wisdom and insight he can on the subject. He'll think long and hard and consider every angle. And no one – I mean no one – can hold a silence like Mark. I'm a talker – if there's so much as a moment of silence in a conversation, I'll burble on witlessly to fill the void. Mark will let the silence hang in the air as he muses, often to an excruciating degree. But when he does open his mouth, the response is invariably thoughtful and considered. He may be perceived as the epitome of the engineer's mindset – all logic, no emotion – but that shouldn't be mistaken for a lack of depth or curiosity. Quite the opposite.

At the same time, he may be the most competitive person I've ever met. And I say that as a former frontline politician and a product of the British private school system, so I've met some pretty competitive people in my time.

One of the ways this competitive streak expresses itself is through Mark's love of mixed martial arts (MMA). He is a big fan of UFC and takes his own training extremely seriously – so much so that his MMA fighting had to be disclosed to investors as a potential business risk. He may have emerged into the public consciousness two decades ago as a skinny nerd, but he's certainly bulked up since then. Just ask Elon Musk, who challenged Mark to an MMA fight in an exchange on Twitter, and then made all manner of excuses to avoid actually getting in the cage with him. Mark's commitment to MMA is so strong that he insisted one morning, during a management offsite, that some of his most senior executives join him for a training session at his purpose-built gym. We all paired off to practise some moves under the watchful eyes of Mark's professional instructors, which meant I found myself wrestling with my then deputy, Joel Kaplan. At one point this involved a manoeuvre apparently known as the 'Domination Mount', in which Joel straddled me and we grappled awkwardly in a way that was, let's just say, a little too close for comfort. It was corporate bonding taken to a whole new level. Joel later jokingly confessed that he had considered reporting it to our then head of HR, Lori Goler, but when he looked up to find her, he saw that she had Mark Zuckerberg in a chokehold. At least surviving such an ordeal meant that Joel was battle hardened, if somewhat bizarrely, to meet the challenge of succeeding me as head of the company's Global Affairs operation when I left Meta some time later.

While you certainly won't find Sheryl Sandberg in an MMA ring, in another life she'd have been a brilliant politician. In the endless world of data, metrics and slide-deck presentations that guides decision-making at Meta, Sheryl brought a no-nonsense ability to cut straight to the heart of the matter, especially as it related to the business. Her commercial nous, and business and political savvy, made her a perfect complement to Mark's remorseless focus on innovative technology and building the best products.

When Sheryl joined Facebook as Chief Operating Officer in 2008, it made $272 million in revenue. In 2021, the last full year before she stood down from the role, its revenue was $118 billion – a more than 43,000 per cent increase. She helped build the company into an advertising behemoth and professionalise its operation. She was also a trailblazer for women in Silicon Valley – both through her example as a strong, unapologetic female leader in a world of bicep-flexing tech bros, and through her vocal advocacy for women in leadership positions and in the world of work more broadly.

When Sheryl first called me about working at Facebook, I was halfway up a mountain on a hiking trip in the Alps. Maybe it was the altitude, but I dismissed the idea out of hand. I couldn't see myself living and working in Silicon Valley, and I couldn't see how Miriam and I could uproot our three school-age boys, the oldest of whom had recently recovered from cancer and was preparing for his A levels. Nonetheless, I sent a short paper to Mark and Sheryl setting out my observations about Facebook's political, policy and reputational challenges and some ideas on how it should change its approach – from a changed posture towards governments and regulators to greater transparency in its products – but I wasn't convinced they would be interested in what I was proposing. I was wrong on both fronts. The paper was evidently important in persuading Mark and Sheryl that I was the right person for the job, and their enthusiasm persuaded me that they were prepared to take change seriously.

Having initially rebuffed Sheryl's approach, and after many phone calls in which she persuasively put the case, I visited California to meet her and Mark. And I'm glad I did. Despite my earlier misgivings about the job, they convinced me that they were both aware of the unique challenges the company faced at that time – in the wake of the Russian interference in the 2016 US Presidential election and the roiling controversy around Cambridge Analytica – and were prepared to make significant changes to move forward. Being part of that was too great a temptation, as was the idea to Miriam and me of a spell in the bright California sunshine after the joyless convulsions in Britain about Brexit. It helped that we had taken a family vacation in Northern California a couple of years earlier, so our three boys were fired up by the memories of a holiday adventure. And so – after a mad dash to find school places for the boys, and a house to live in – Miriam and I decided to take the plunge and I started working at Facebook in October 2018, and we moved to a new family life in Silicon Valley shortly afterwards.

What attracted me to the role at Facebook was three-fold. First, I am an optimist about technology. Technology – from medicine to transport, from communication to computing – has been at the heart of the remarkable rise in living standards for billions of people over the last two centuries. I believe wholeheartedly that the age of rapid technological progress we are living through today has the potential to continue to change the world and the lives of its citizens for the better. The remarkable advances we are seeing now in AI, virtual and augmented reality, robotics, biotech and more have the potential to help us all live longer, safer, more comfortable lives. And I believe they will help us to rise to the major challenges that face us as a planet.

That doesn't mean I'm an uncritical 'techno-optimist', a term that has taken on a politically loaded association with a particularly libertarian strand of Silicon Valley idealism. Nor am I some kind of belligerent absolutist who thinks any form of regulation or government intervention is stifling the full-throated roar of technological progress. Far from it. I believe technology should serve society and not the other way round. And society – through politics – is right to exercise caution and scrutiny in the face of rapid change, uncertainty and shifts in power. I just don't think there is anything mutually exclusive between being optimistic about technology and pragmatic about ensuring innovation happens within democratically accountable guardrails.

Second, the interactions I had with tech companies both during and after my time in government opened my eyes to just how profound the issues they were dealing with were. These were companies with a borderless, global reach, and they were necessarily international in their outlook. The issues their products raised were at times ethical and philosophical, and at times technical or commercial. Above all, they were consequential. Their products were part of the everyday lives of billions of people. And they were constantly evolving.

Third, I was convinced of the need to find a way, somehow, to build bridges between the worlds of technology and politics. It wasn't that I considered myself in any way an expert at doing that – and it wasn't like I had a blueprint for how to do it – but it was a challenge I found both necessary and fascinating.

When I arrived at Facebook in 2018, I found a company still largely in shellshock. Many of the employees had joined when the company's reputation was starkly different: an exciting, idealistic place led by a boy genius, making cool things that people were taking up at a phenomenal rate, generating oodles of cash, and with a high-minded mission to connect the world. Going from whizz-kids to public enemy number one was a culture shock few were truly prepared for. It reminded me a little of the Liberal Democrats during the coalition – a collection of idealists who wanted to change the world and who couldn't quite figure out why people saw them as the bad guys.

An internal culture change was already under way. The famous 'Move Fast and Break Things' slogan had been jettisoned in favour of 'building responsibly'. There was a significantly greater focus on what the tech industry refers to as 'integrity' – ensuring the robustness of security systems and the safeguards in place to protect users, proactively detecting and removing content that violates content policies, creating an industry-first network of international fact-checking partners, tackling foreign and domestic attempts to spread disinformation on its platforms, and greater cross-industry intelligence sharing and cooperation to take on malicious groups operating across platforms.

Today, it's a very different company indeed. What started as a social media company has expanded its technological focus to AI and 'metaverse' technologies like virtual and augmented reality, a shift that had been under way behind the scenes since long before I joined. And the renewed internal focus on responsibility has turned it into an industry leader in safety, transparency and governance. That's not to say the idealism about technology has drained away. Nor has the thirst for technological innovation been tempered. It still makes mistakes, and when a company with this scale and reach makes mistakes, they have real consequences. But there has undoubtedly been a mindset shift. It's a company much more aware of the social responsibilities that come with its size and influence, much more aware that its actions (or inactions) have real-world consequences, and much more sophisticated in understanding that its long-term self-interest rests in engaging with the worlds of politics, civil society and academia constructively and not defensively.

So much of the debate about technology these days has been reduced to pantomime baddies, knee-jerk assumptions and sensational headlines. Often it is the company that I worked for that is being caricatured and criticised. You may well feel the arguments I make in this book are an attempt to defend the honour of my former corporate paymasters. In some cases they are, though only when I believe there is honour to defend. But my intention in writing this book is not to mount some great defence of Meta. My role gave me a unique vantage point at the intersection of technology and politics. It is a company that is at the sharp end of world-changing events. Operating the world's biggest social networks has made it immensely wealthy and immensely controversial, a lightning rod for the techlash in a way that perhaps no other internet platform has been. It is also a global company benefiting from the borderless global internet, with billions of people in most nations on earth using its services at a time when governments are trying to reassert their national sovereignty. And it is one of a small handful of powerful tech companies investing billions of dollars in foundational AI infrastructure and research in order to try and lead the coming technological wave. This book is all about power – and Meta undoubtedly has it. I hope that the observations I am able to make about the relationship between technology and politics are insightful because of this unique vantage point, rather than undermined by it.

In the years since I joined Meta, I've had little glimpses of the future – from smart glasses that translate text and audio in real time like Douglas Adams's Babel Fish, to photorealistic avatars that allow you to talk to someone a thousand miles away as if you are in the same room. I've been immersed in the invigoratingly optimistic, if sometimes naive, world of the innovators and engineers building these staggering new technologies. And I've been on the front lines of the techlash, confronted by critics and lambasted by politicians in both public and private. I've been the decision maker internally in the company on some of the most controversial decisions about what content to keep up, and what content to take down, and I've acted as a kind of diplomat on its behalf externally, making its case to world leaders, regulators and policymakers. Throughout, I tried to build bridges between the worlds of politics and technology. I tried to steer the company as best I could to a place where it interacts with the political world in a more mature, transparent and accountable way, where it is able to demonstrate not only the benefits of the remarkable technologies it creates, but also the lengths it goes to in order to function in a way that understands and accepts the social responsibilities inherent in building technology utilised at such scale.

But the real purpose of this book is not to defend myself or Meta or Big Tech. It is to raise the alarm about what I believe are the truly profound stakes for the future of the internet and for who gets to benefit from these revolutionary new technologies. In the chapters that follow, I will draw on my atypical confluence of experience in both politics and technology in order to set out, to the best of my ability, how and why we have arrived at this pivotal moment for the internet, as well as what we might do about it. And to do this, I must begin by examining the fears around the key technologies of social media and AI and to set out, candidly and based on evidence, where these concerns are justified, where they are based on misconceptions or mistaken assumptions, and where expert opinion is genuinely divided.

PART ONE

The Power of the Platforms

CHAPTER 1

The Trouble with Social Media

Facebook has been depicted as a 'doomsday machine', designed, in the words of Jason Pontin, former editor-in-chief of *MIT Technology Review*, to 'distract, divide, and madden' and leading to the dissolution of 'our social, civic, and political ligands'. Sociologist Zeynep Tufekci said YouTube 'may be one of the most powerful radicalizing instruments of the 21st century'. And in a 2022 article in *The Atlantic*, social psychologist Jonathan Haidt argued that the ability to share and retweet content on Facebook and Twitter has made Americans 'stupider'.

Whether you share the outrage of these academics and commentators or not, the chances are you share some of their assumptions. The idea that social media has had a negative impact on democracy, that it has made us more polarised and angry, that it has stranded us in echo chambers where we only see and hear from people and news reports that confirm our own world view and our own political biases, that it has given populists a megaphone, that the opaque business models of social media companies incentivise outrage and extreme content, and that social media has left us vulnerable to being manipulated by malign foreign interference – all of this is now accepted wisdom in many quarters. By this logic, the fact that social media has proliferated at more or less the same time as a particularly turbulent political era – compared with the relatively calm period of domestic politics experienced in the 1990s and 2000s by those of us in western liberal democracies – is no coincidence.

Indeed, it is true that the two phenomena have occurred in tandem; social media is of course used by many people to express their political views loudly and emphatically; politicians and their supporters use social media very prominently to campaign; and there have been a number of high-profile scandals – like Cambridge Analytica and Russian attempts to interfere in the 2016 US Presidential election – in which these tools have been exploited for nefarious ends by people who want to influence the outcome of elections. There is undoubtedly a relationship of some kind between social media and today's political volatility. But there's also a lot of putting two and two together and getting five.

Many of the critiques of social media start with the assertion that the fundamental problem lies with the business model of social media companies which, in the critics' telling, is reliant on keeping users scrolling endlessly on their phones so as to feed them as many advertisements as possible. To keep people glued to their feeds, so the argument goes, apps like Facebook, Instagram, TikTok and Twitter/X are specifically curated to 'engage' users – in other words, to interact with the content by liking, commenting on or resharing it – and the content that engages people the most is the stuff that gets them riled up. So social media companies feed them more and more emotionally charged content, pushing people into a vicious cycle whereby they consume ever more extreme content and sink deeper and deeper into echo chambers where the only people they encounter are those who agree with them and the only news they consume is that which reinforces their world view. People get angrier. Political views harden. Conspiracy theories abound. Objective truth disappears. Trust in institutions nosedives. Society gets increasingly divided.

This is the essential narrative of 'Surveillance Capitalism', a phrase and an influential thesis popularised by the Harvard academic Shoshana Zuboff. In her explanation of platform economics,

Zuboff asserts that digital platforms have cracked the code of how to predict human behaviour and sell that prediction to advertisers. Digital platforms employ, she notes, 'automated machine processes' that 'not only know our behavior but also shape our behavior at scale'. The idea that digital platforms – or anyone, really – can know not only what people think but also what they will think in the future is a strong claim to make. Even in an otherwise sympathetic review of Zuboff's book, *The Age of Surveillance Capitalism*, media scholar Ella Hafermalz notes that:

> Zuboff argues that the real profit in Surveillance Capitalism comes not only from predicting our behavior, but also in *modifying* it [...] The argument seems to be that the best way to predict behavior, is to control it [...] This is however hard to comprehend as a business model. Who is the producer? Who is the consumer? And what role are different intermediaries playing? What is the 'original' behavior, versus the 'modified' behavior, and which is being sold in the form of prediction profits? For example, did I really choose to review this book? Or am I the victim of behavior modification engineered by Twitter, Google, and Amazon on behalf of their clients?

Surveillance Capitalism is a compelling and potentially alarming theory. But it isn't true. To understand why, it's worth looking at where the incentives really lie in the business model of companies like Meta. When it comes down to it, Meta's business model is less mysterious than is sometimes alleged: it makes its money through advertising. Meta doesn't sell people's data. Advertisers – from giant corporations to millions of small- and medium-sized businesses – tell Meta the sort of people they are trying to reach. In some cases they may already know the specific people they want to advertise to – for example, existing customers whose details they hold, or people who have signed up to a mailing list – but in most cases they are aiming to reach categories of people rather than known individuals. Meta's advertising tools offer them demographic, location-based or interest categories to choose from, then they show the ad they have provided to people who best fit their description. The price the advertisers pay is set by an auction, ensuring the price is always the lowest it can be in the category in which advertisers are competing to reach the same demographic. Meta doesn't – and can't – artificially jack up the prices. The auction also takes into account the quality of the ad and a prediction of how likely someone is to act on it (for example, by clicking on it) when selecting a winner, helping to ensure that users see good, relevant ads.

That's it. It's really that simple. A number of tools – including ones using AI – exist to help those advertisers refine the audiences they reach, identify alternative audiences the ad may appeal to, test the performance of their ads against these audiences, and even generate or refine the content they use in the ads themselves. The targeting of audiences that is possible on Facebook and Instagram is undoubtedly more sophisticated than anything that has come before, but it is essentially the same activity that has defined advertising since the industry was born. At the end of the day, advertisers are still just doing what they have always sought to do – get their adverts in front of the people most likely to buy their products.

Of course, if you think trying to persuade people to buy things is inherently bad, then you may find yourself in sympathy with Zuboff's argument. Her thesis doesn't so much critique the merits and pitfalls of the value proposition that digital platforms present to users; rather it takes for granted that any form of capitalism is problematic by default. *The Social Dilemma* is a Netflix documentary that brings Zuboff's theory to life with emotive fictitious scenarios (including a sinister digital control room where engineers press buttons and turn dials to manipulate a teenage boy through his smartphone) in order to build an argument that the targeting of ads on social media is uniquely manipulative. In the film, social media critic Tristan Harris claims that apps like Facebook have 'hijacked our minds', arguing that, by understanding you better, the companies behind these platforms can persuade you to do, think and buy things you wouldn't otherwise, as if you're some kind of passive stooge meekly susceptible to being brainwashed.

It's deeply patronising. We all see adverts every day. When we see things we like the look of, if the ads are persuasive and well timed, we might buy them. When we see things we don't fancy, we don't buy them. People have agency. The tools available to advertisers make it more likely than ever before that you will see more ads for things you like than for things you don't. But they're still just ads.

Advertising is a sophisticated and creative industry, built on decades of detailed analysis and understanding of consumer sentiment and behaviour, and often informed by theories of human behaviour and psychology such as behavioural economics. Brilliant marketers can create effective and persuasive ad campaigns – and they can test and target them in a much more granular way online. But the principles are the same whatever the medium. It's not a fundamentally different and more sinister exercise just because it takes place on social media.

As Mike Godwin (the inventor of 'Godwin's Law', which holds that the longer a conversation on the internet goes on, the more likely it becomes that someone will be compared to a Nazi) put it in a post, ironically on Facebook:

> My big question for those who believe Facebook has overcome the free will of 2 billion people: how did all of you escape? [. . .] Tristan Harris's argument, which is weird to me, is that, when you encounter targeted ads or messages in social media, you will have thoughts put into your brain that you may not have wanted to have, and that you may not wish to have. My response is, dude, have you ever had a conversation before?

Ads-based business models have one big advantage for consumers. Because companies like Meta make their money selling personalised ads, they don't need to charge users to use Facebook or Instagram or WhatsApp or Messenger. And they also happen to be good for small businesses who can't afford to reach customers with expensive television or billboard campaigns but can afford to pay for much cheaper digital ads targeted at clearly defined and often local audiences that might be interested in what they're selling. Of course, some see this as sinister. The old adage 'If you don't pay for the product, you are the product' gets repeated a lot these days. But you are no more the product in the case of social media than you are in any advertising-based industry.

But what about the incentive to keep you engaged with ever more extreme content? Again, it's largely accepted wisdom that this is what social media companies do. But it's worth thinking about their incentives. If your business is selling ad space, then you need to be able to persuade advertisers not just that they will see a strong return for the money they spend, but that you are providing a space where they want their brands and products to appear. Companies care about their reputations, and most don't want to become associated with controversy. They certainly don't want their products placed next to hateful or extreme posts, or in between hateful or extreme videos. In the wake of the decision not to remove President Trump's controversial 'when the looting starts, the shooting starts' post in 2020, more than a thousand advertisers joined a boycott of Facebook organised by the #StopHateForProfit campaign – including some of the company's biggest advertisers like Pfizer, Ford, Coca-Cola, Best Buy, Adidas and Starbucks. Similarly, big-brand advertisers abandoned Twitter/X after Elon Musk took the company over, fearing his lax approach to tackling hate speech. Social media companies make far the largest part of their revenue from advertising – and a large chunk of that from household-name brands with big marketing budgets. Scaring them away simply isn't in their financial interest.

Then there's the question of whether social media companies really are trying to keep you doom scrolling for as long as possible. Of course, it is in Meta's short-term interest if people use its apps for longer and spend that time liking, commenting on and sharing content. But how long you spend on the app in a single session is not nearly as important as getting you to come back over and over again. Repeat business is good business. The company's focus since day one has been on growing its user base – that means getting more and more people using the apps and, crucially, retaining the users they already have. Meta's primary internal measures of success are DAU (daily active users) and MAU (monthly active users) – there's always an acronym! – and those numbers go down if users stop returning to the apps. When people's experiences using Facebook or Instagram are full of negative emotions, they may be more likely to engage with the content they see, and as a result they may well spend more time on the app in the short term. But it's not what keeps them coming back day after day, month after month. For that to happen, they have to find the experience useful and meaningful – whether it's keeping up with old college friends, seeing pictures of their nephews and nieces, watching videos of things that interest them or finding out what people are complaining about in their neighbourhood. That's why Facebook's algorithmic ranking systems take into account far more than whether a piece of content is likely to be engaged with – the intention is to show you content that is *meaningful* to you, not simply to give you a momentary dopamine spike.

The truth is most people don't really use social media to engage in politics. Politicians and newspaper columnists assume that what they see when they scroll through their feeds is what everyone sees. It isn't. Most of the content people see in their Facebook feed, even in an election season, is not about politics at all. Political content makes up a very small part of the content people see on Facebook – less than 6 per cent. Most people still get most of their news from television. Of course that may not apply to people who work in politics or journalism, whose interests inevitably lead them to consume much more political and news content. And politically engaged people tend to be connected to other politically engaged people – invariably on Twitter/X, which is awash with people yelling at each other about politics, not on Facebook or Instagram – so it follows that they are more likely to see a higher proportion of political content shared by people in their networks. As a result, the experience of those who form the narrative around social media is skewed in a way that is unrepresentative of most people's experiences.

The biggest misunderstanding of the relationship between people and the algorithms that serve them personalised content is that we are somehow passive recipients, accepting uncritically the barrage of content thrown at us by complex systems designed to manipulate us and sell us things. This idea misunderstands the purpose of these systems in the first place. In order for you to spend more time or money on a service, whether it's a streaming service, an e-commerce platform or a social media app, the algorithms need to be responsive to you. Your tastes. Your behaviour. Your interests. Meaning that, far from you passively receiving content, you are required to actively participate in the process. The very idea of personalisation – the driving force behind so much internet innovation in the last couple of decades – suggests an active engagement by the individual user with the digital platform so that the former can enjoy relevant experiences delivered by the latter.

In an article I wrote in 2021, I used the analogy of a couple planning dinner. Imagine you're on your way home from work when you get a call from your partner. They tell you the fridge is empty and ask you to pick some things up on your way home. You choose the ingredients, they cook dinner. So you swing by the supermarket and fill a basket with a dozen items. Of course, you only choose things you'd be happy to eat – maybe you choose pasta but not rice, tomatoes but not mushrooms. When you get home, you unpack the bag in the kitchen and your partner gets on with the cooking – deciding what meal to make, which of the ingredients to use, and in what amounts. When you sit down to eat, the dinner in front of you is the product of a joint effort: your decisions at the grocery store and your partner's in the kitchen. Neither of you is passive. You both have agency.

No analogy is perfect, and this one shouldn't be taken literally, but ultimately content ranking is a very similar dynamic partnership, in this case between the user and the algorithms. The final product – what you see in your social feed, the movies and television shows you are recommended by a streaming service, or the products you are presented with as possible purchases on an e-commerce platform – is not foisted upon you, but has been shaped at every stage by your own tastes, behaviours and interests.

One significant development since I first used that analogy is the rise of TikTok and the algorithmic sharing of 'unconnected content', meaning content that has not been shared by your friends or people you have actively chosen to follow. Meta has developed its own 'discovery engine' model to serve a similar purpose on Facebook and Instagram. But these systems too, while being further removed from your direct social connections, are still based on the principle of learning from your preferences in order to serve you content you will enjoy or find meaningful.

What's more, Facebook and Instagram are far from a kind of viral free-for-all. Meta has a detailed rulebook – what it calls its 'Community Standards' – dictating what content is and isn't allowed on Facebook and Instagram, and invests billions of dollars in systems and people to identify and enforce these policies, removing or limiting the distribution of problematic content. Indeed, the AI systems it uses to identify and remove hateful content have improved rapidly in recent years. Meta also partners with a global network of independent fact-checking organisations working across sixty languages to flag false or partially false content, which is then labelled.

But how and to what extent Meta moderates content is constantly evolving. In 2025, Meta announced its most significant changes to its approach to content since 2016. It is going to be moving away from independent fact-checkers, starting in the United States, and towards a 'community notes' style approach, whereby a community of contributing users decide when posts are potentially misleading and need more context, and people across a broad range of perspectives agree what context users should see, all of which is in keeping with a wider shift in the industry to user-based systems. It is also narrowing the scope of how it applies its automated systems to catch bad content – focusing on illegal and severe policy violations like terrorism, child sexual exploitation, drugs, fraud and scams, and relying more on users to report less severe violations. The truth is, even the most sophisticated systems will make mistakes, and with countless millions of pieces of content being posted every day, even a small proportion of mistakes amounts to action being taken against millions of pieces of harmless or trivial content. For example, in December 2024, the company estimates that of the millions of pieces of content it removed every day, one to two out of every ten of these actions may have been taken against content that didn't actually violate its policies.

There's certainly room for legitimate debate about whether Meta's policies, or those of any other social media company, draw the line in the right place about what is and is not allowed on their services, and whether it should even be up to them to draw these lines in the first place. But if it were true, as is frequently alleged, that social media companies have an incentive to keep users glued to their apps by showing them ever more emotive content, you would expect whatever content-moderation rules they have to be opaque and designed as loosely as possible, and that they would be enforced weakly, if at all, and that the company would make a merely token investment in the teams and systems that enforce them. None of that can reasonably be said of Meta's approach (though arguably it can for Twitter/X following Musk's takeover). Journalists and activists will continue to keep the pressure up on companies like Meta to be even more transparent and accountable, and rightly so. But any meaningful scrutiny of Meta's policies and practices will find a company that is doing far more to keep bad content from spreading across its services than it is obliged to.

Has social media destroyed democracy?

At the heart of the criticisms made by those like Shoshana Zuboff and Tristan Harris is the assertion that social media is a driving force behind the rise in polarisation in the United States – and, by extension, the rest of the world – and therefore a causal factor in the rise of anti-establishment populist leaders and movements, from Donald Trump and Argentina's Javier Milei to Brexit. But political polarisation and the causes behind it have been the subject of swathes of academic research in recent years, and there isn't a great deal of consensus. In fact, many studies suggest that social media is not the primary driver of polarisation at all.

Research from Stanford looked in depth at trends of what academics call 'affective' polarisation in nine countries over a period of forty years. Affective polarisation refers to the 'extent to which citizens feel more negatively toward other political parties than toward their own'. Affective polarisation can be contrasted with 'ideological' polarisation, which refers to a disagreement on the basis of political opinion – a disagreement on political issues rather than an emotional dislike of the people who disagree with our political beliefs. The Stanford research found that in some countries affective polarisation was growing slightly before Facebook even existed, and in others it has been decreasing while internet and Facebook use have increased. The country that had the greatest rise in polarisation during that period was the United States. Similarly, an Oxford University literature review of studies in this field found that polarisation has fluctuated internationally – in some countries it has increased, while in others it has decreased. As the authors note: 'levels of affective polarisation vary greatly by country (complicating the notion that polarisation is pronounced everywhere)'.

Professor Axel Bruns of Queensland University of Technology in Australia wrote a prescient book back in 2019 called *Are Filter Bubbles Real?* In it, Bruns argued that the widespread tendency to blame platforms and their algorithms for political disruptions obscures far more serious issues pertaining to the rise of populism and hyperpolarisation in democracies. Evaluating the evidence for and against echo chambers and filter bubbles, Bruns offered a persuasive argument for why we should shift our focus to other causes of polarisation, including 'socioeconomic inequalities, citizen disenfranchisement, and inflammatory political propaganda'.

Data published in the EU suggests a similar finding. For example, one study looked at the European Union's Eurobarometer – one of the most comprehensive Europe-wide public opinion surveys – to better understand whether those people who get their news primarily from social media hold less politically diverse attitudes than those who rely on more traditional media such as radio, television, printed newspapers and non-social media internet sites. The results provide little evidence to support the 'echo chamber' effect of social media. In fact, they suggest that levels of ideological polarisation are similar whether you get your news from social media or elsewhere.

As a European, it is frustrating that so much of this debate is US-centric. More than 90 per cent of Facebook's users are in the world beyond America's borders. The US is far from the only country experiencing a rise in political polarisation, but there are many places where this isn't the case – and they have social media too. Even within America, the connection between social media and polarisation is unclear. A Harvard study published ahead of the 2020 US Presidential election found the disinformation campaign that led many to believe election fraud is associated with mail-in voting was driven primarily through elites and mass media. The authors noted that their results were consistent with a previous study looking at 2015–2018, which found that Fox News and the Trump campaign were 'far more influential in spreading false beliefs than Russian trolls or Facebook clickbait artists'. They said that this trend was even more pronounced in the 2020 election cycle, 'likely because Donald Trump's position as president and his leadership of the Republican Party allow him to operate directly through political and media elites, rather than relying on online media as he did when he sought to advance his then-still-insurgent positions in 2015 and the first half of 2016'.

And research from both the Reuters Institute in 2017 and

Pew in 2019 showed that you're likely to encounter a more diverse set of opinions and ideas using social media than if you only engage with other types of media.

Ahead of the US Presidential election in 2020, Facebook gave access to a team of external academics to conduct research to better understand the impact of Facebook and Instagram on key political attitudes and behaviours during that election cycle. The team was led by Professor Talia Jomini Stroud, founder and current director of the Center for Media Engagement at the University of Texas at Austin, and Professor Joshua A. Tucker, co-director of the Center for Social Media and Politics at New York University. Professors Stroud and Tucker selected fifteen additional researchers to collaborate on the project. Facebook's internal researchers designed the studies together with these external partners. Importantly, neither Facebook's researchers nor the company as a whole had authority to restrict the study's findings.

The researchers wanted to get a better understanding of how social media algorithms impact people's political beliefs. They looked, for example, into whether algorithms lead to echo chambers and whether they polarise people; what the impact of viral political information was on political beliefs; and whether switching to a chronological feed – effectively switching off the algorithm – had an impact on people's political behaviour.

Most of the studies were social science experiments, which is a method that is considered the gold standard in university research, especially when the goal of the research is to understand how a certain variable (in this case the algorithm) might causally affect other variables (like political polarisation). The experimental findings add to a growing body of research showing there is little evidence that key features of Meta's platforms alone cause harmful affective polarisation. They also challenge the now commonplace assertion that being able to reshare content on social media drives polarisation. For example, one paper published in the scientific journal *Nature* states that 'these findings challenge popular narratives blaming social media echo chambers for the problems of contemporary American democracy'. And the co-chairs of the study have stated: 'Removing reshared content on Facebook decreased news knowledge among the study participants, and did not significantly affect political polarization or other individual-level political attitudes.'

One of the papers showed there was considerable ideological segregation in consumption of political news, reflecting a complex interaction between algorithmic and social factors. Yet when participants in the experiments saw less content from sources that reinforced their views, they were actually more likely to engage with it. But even then it had no discernible impact on polarisation, political attitudes or beliefs.

Of course, critics will no doubt say that I have cherry-picked a handful of studies and findings that suggest social media isn't the root cause of polarisation, and they will point to others that say the opposite is true. And they'd have a case. Like I said, the research is mixed. There are credible studies that suggest that leaving Facebook for a month reduced political polarisation among those studied, or that exposure to opposing views on social media (the opposite of the filter bubble idea) can increase political polarisation.

According to Joshua Tucker and Stanford scholar Nate Persily, those who study social media and democracy tend to fall into two camps:

> The first emphasizes the rise of social media echo chambers, fake news, hate speech, 'computational propaganda', authoritarian governments' online targeting of opponents, threats to journalism, and foreign election interference. The other school challenges the independent significance of the shift in technology (as opposed to other sociological factors) while also suggesting that the magnitude and prevalence of the alleged technology-related problems are overblown.

You can be in one camp or the other – as you'd expect, I'm sympathetic to the latter – but what you certainly can't claim is that there is consensus. The conclusion I'd encourage you to draw is not that there is no relationship between social media and polarisation. It is simply that it's not clear if there is a relationship or what it is, and therefore the accepted wisdom that social media makes us more polarised is far from established fact.

So, has social media destroyed democracy? Of course not. Perhaps the better question is 'How has social media *changed* democracy?' Because undoubtedly it has. Like other communications technologies before it – from the printing press and the radio to the television and the internet – it has democratised speech, making it possible for previously disempowered people to make themselves heard, in turn weakening the power of the gatekeepers and institutions who until then held a tight grip on the flow of public information. That's a disruptive and messy change. It brings with it huge opportunities and it comes with unexpected and unintended consequences, as well as both practical and philosophical challenges. It will take time to fully understand the implications of new technologies on society. But we should not lose faith in the idea that empowering people is both a good thing in itself and a long-term benefit for our societies.

Teens and mental health

If the impact of social media on politics and public discourse – whether that impact is real or assumed – is cause for angst-ridden debate, it is nothing compared with the fear around the impact of new technologies on the mental health of children and young people. We are frequently told that high smartphone and social media use can be hugely damaging to teenagers, amplifying peer pressure and heightening social anxiety, and in some tragic cases contributing to self-harm and even suicide. Nothing causes more fear than the idea that our children could be coming to harm because of forces we don't understand and can't control. The idea that this is the result of technologies created by companies that are perceived to put profit above duty of care to our children is both disturbing and infuriating. Facing sincere and emotive criticism of this kind has been the hardest part of my job representing one such company. My three boys have grown from children to young adults during the last decade, and Miriam and I have wrestled with all the dilemmas and anxieties that come with raising smartphone-native teenagers living so much of their lives online. I can't dismiss the fears other parents feel, nor can I dismiss parents' desire for reassurance that these technologies are safe.

These fears have been captured most effectively in Jonathan Haidt's book *The Anxious Generation*, in which he makes the case that smartphones and social media use are directly linked to a decline in the mental health of teens and young people:

> Once young people began carrying the entire internet in their pockets, available to them day and night, it altered their daily experiences and developmental pathways across the board. Friendship, dating, sexuality, exercise, sleep, academics, politics, family dynamics, identity – all were affected. Life changed rapidly for younger children, too, as they began to get access to their parents' smartphones and, later, got their own iPads, laptops, and even smartphones during elementary school.

Even though Haidt is articulating a theory that feels intuitively true for many of us, his assertions are far from accepted fact among academics who research these issues. An international study which looked at more than 900,000 adolescents across thirty-six countries showed no change in life satisfaction between 2002 and 2018. Max Roser, of Our World in Data, looked at suicide rates among older teenagers and young adults across a number of European countries and found they have largely held steady or declined between 2000 and 2019 in France, Spain, Italy, Austria, Germany, Greece, Poland, Norway and Belgium, with only very small increases in Sweden. World Health

Organization data shows that teen suicides in many European countries declined in the same period, and that suicide rates in the US among 15–19-year-old boys were actually higher in 1990 than today. Haidt also ignores evidence that in the US recorded rates of teenage suicide and depression have increased because of changes in classification for insurance purposes, and dismisses the idea that society's increasing acceptance of mental health disorders may partially explain the rise in reporting of teenage mental health issues.

In response to Haidt's book, psychologist Candice Odgers, a professor at the University of California, Irvine, contributed an article to *Nature* noting that:

> Hundreds of researchers, myself included, have searched for the kind of large effects suggested by Haidt. Our efforts have produced a mix of no, small and mixed associations. Most data are correlative. When associations over time are found, they suggest not that social-media use predicts or causes depression, but that young people who already have mental-health problems use such platforms more often or in different ways from their healthy peers.

Odgers then doubled down on her rebuttal of Haidt's thesis with an article in *The Atlantic*, with this observation for parents:

> We should not send the message to families – and to teens – that social-media use, which is common among adolescents and helpful in many cases, is inherently damaging, shameful, and harmful. It's not. What my fellow researchers and I see when we connect with adolescents is young people going online to do regular adolescent stuff. They connect with peers from their offline life, consume music and media, and play games with friends.

As *New York Times* columnist David Wallace-Wells puts it:

Over the past five years, 'Is it the phones?' has become 'It's probably the phones,' particularly among an anxious older generation processing bleak-looking charts of teenage mental health on social media as they are scrolling on their own phones. But however much we may think we know about how corrosive screen time is to mental health, the data looks murkier and more ambiguous than the headlines suggest – or than our own private anxieties, as parents and smartphone addicts, seem to tell us.

In 2021, a former Facebook product manager called Frances Haugen leaked reams of internal documents to the media, including slides from an internal research deck about Instagram's relationship with mental health among teenage girls. For all the sensational headlines that followed – for example, the *Wall Street Journal* declared 'Facebook knows Instagram is toxic for teen girls, company documents show' – the research in question was far less declarative. The slides broke down the research findings into twelve areas of mental health and wellbeing, including loneliness, anxiety, sadness and eating issues. In eleven areas out of the twelve, the majority of teenage girls who said they struggled with that particular issue also said Instagram made those difficult times better rather than worse. Contrary to the headlines, the research did not conclude that Instagram was inherently bad for teenage girls overall; like much external research, it found that many troubled teenagers find comfort and support through social media. However, it did also show that, for a sub-section of vulnerable teens, that wasn't the case. And that finding should not and must not be ignored.

The difficulty for anyone trying to navigate between claim and counterclaim over social media's effect on teenage mental health is that, as with polarisation, the evidence is genuinely mixed and certainly does not suggest there is a clear population-wide problem. There are important nuances too: how involved parents are in a child's online life can make a meaningful difference to the child's wellbeing; for many young people, social media is a net positive for their wellbeing, providing an outlet for self-expression and an ability to connect with friends and discover others who share their interests; and this is especially the case for many young people who are otherwise marginalised in their local communities – for example, LGBTQ+ teens who can find role models, communities and, most importantly, acceptance online that they can't find where they live.

The leading digital media researcher danah boyd has argued that the simplistic advice simply to 'disconnect' smartphones or social media is not a panacea for solving the mental health crisis:

> As a researcher, I know that young people's relationship with tech is so much more complicated than pundits wish to suggest. I also know that the hardest part of being a parent is helping a child develop a range of social, emotional, and cognitive capacities so that they can be independent. And I know that excluding them from public life or telling them that they should be blocked from what adults value because their brains aren't formed yet is a type of coddling that is outright destructive. And it backfires every time.

Experimental psychologist Amy Orben, working at the University of Cambridge, has found that social media wellbeing may vary based on age and gender: 'a decrease in life satisfaction ratings' is associated with high social media use 'for males (14–15 and 19 years old) and for females (11–13 and 19 years old)'. In line with previous studies, Orben's research finds that these associations are small in statistical strength, but also notes that the effects are likely to 'differ across individuals, as each person's sensitivity is further influenced by a wide range of individual, peer, and environmental dynamics'.

No one – even the most ardent defender of social media, or critic of Haidt – is ever going to deny that for some teens, some of the time, the online experience is not good. Even if social media use is broadly positive and empowering for a large majority of young people, there will still be a minority of vulnerable young people for whom it can be damaging. Common sense dictates that we proceed cautiously and cater as much as we can for those young people who may need more guidance and vigilance in how they interact with content online.

While I was at Facebook/Meta, I consistently made the case for more parental controls and safety features in our products, more resources for the teams that work on these issues within the company, and more research into the impact of social media on young people's wellbeing. Today, Facebook and Instagram have more controls available to parents and young people, more policies designed to keep them safe, and significantly more resources devoted to those controls and policies than when I joined the company.

In an environment where there are at any moment more pieces of content available on Instagram than books that have been written in the history of humanity, identifying and taking down bad content is a monumental challenge. We can all agree that content that encourages suicide, self-injury, eating disorders or things like bullying and harassment should be taken down, but even with the most advanced technology, enforcement will never be perfect. Meta's teams regularly consult experts in adolescent development, psychology and mental health to help make its platforms safer and age-appropriate for young people, and have introduced a range of parental controls and other features to encourage healthy use and keep teens safe, including letting parents set daily time limits and regular breaks for their children. Such features include 'Quiet Mode', which encourages teens to leave the app and pause notifications if they've been scrolling for just a few minutes at night; a 'Hidden Words' feature to let people filter emojis, words or phrases they don't want to see in comments or DMs; automatically setting the accounts of under-sixteens to private so they won't be contacted by strangers; preventing people over nineteen from sending private messages to teens who don't follow them; and other measures to prevent problematic use and tackle bullying and harassment. In 2024, the company launched Instagram Teen Accounts, with protections like default privacy settings, time limits, safeguards against sensitive types of content, and restrictions on who can message teens and interact with their posts. All of these protections are turned on automatically, with the decision to change settings to be less strict left to parents if teens are under the age of sixteen. And because some teens will inevitably lie about their age to try and game the system, the company has introduced processes to make getting around the controls more difficult, for example by training AI systems to identify these teens and automatically place them in protected settings.

But undoubtedly more must be done. In *The Anxious Generation*, Haidt proposes 'four foundational reforms': no smartphones before high school; no social media before the age of sixteen; phone-free schools; and more childhood play and independence. Writing in response to Haidt's book, psychologist Jacqueline Nesi, an assistant professor at Brown University, expressed scepticism at his findings but sympathy with his proposals:

> This question that we're trying to answer – *Did the introduction of social media and smartphones cause the increase in mental illness among young people in recent years?* – is simply a very hard question to answer with data [. . .] The truth is, it may be a very long time before we have definitive proof one way or the other. So, then, what do we do? [. . .] In general, I agree with some of the solutions Haidt proposes. But, again, I do not think we have strong evidence that they will work. In fact, we will likely never have enough evidence to say that, for example, age 14 is safer than age 13.75 for introducing smartphones, or 16 is better than 15 for social media. So, the question becomes: how much evidence do we need to act?

Like Nesi, I am sceptical of Haidt's methods and arguments, but I actually don't disagree with many of his suggestions. There are a variety of common-sense steps that we as a society could be taking regardless of whether or not there is any direct connection between phone use and mental health outcomes. We need not wait until the scientists and researchers have thrashed it out among themselves, because, even in the face of inconclusive evidence, it is reasonable for the industry to be constantly looking for ways to improve their users' experiences on social media. Far from resisting regulation, I would like to see policymakers agree on the appropriate age limits for smartphone and social media use, and enable companies to implement those limits effectively. Right now, the biggest problem for individual companies in effectively applying age limits and related age-monitoring tools is that there is no easy way to confirm a user's age. But there is an obvious solution: mandate the operating systems (iOS and Android) to share device users' ages when they download apps from the app stores – data the operating systems get as part of the hardware acquisition already. This would be a simple one-step way for parents to control all the different apps that their kids use (in the US, the average teen uses forty different apps per month) and would remedy the fractured app-by-app approach we have today. We should make a societal judgement about whether to set these age limits for smartphones or social media use at thirteen, fourteen, fifteen or sixteen, then write it into law. We should keep smartphones out of schools. And we should certainly find ways to encourage kids to play, interact and be independent away from the alluring glow of the screens. Whether measures like these will reverse the mental health crisis remains to be seen, but that's not a reason not to act now.

We all – parents, companies, governments and individuals – have a responsibility to the young and the vulnerable in our societies. We should not be passive in seeking to protect people who don't have the agency to protect themselves, and we should always question whether we are doing the right thing or whether more can and should be done. For most adults, I believe that, wherever possible, people should be able to choose for themselves, and that people can generally be trusted to know what is best for them and how much risk they are comfortable with.

For children, in most cases, the people who are best placed to make decisions about what they should have access to online are their parents. And they need to be empowered.

But I am also acutely conscious that we need collectively agreed ground rules, both on social media platforms and in society at large, to reduce the likelihood that the choices exercised freely by individuals will lead to collective harms. Politics is in large part a conversation about how we define those ground rules in a way that enjoys the widest possible legitimacy, and the challenge that social media now faces is, for better or worse, inherently political.

Should a private company be intervening to shape the ideas that flow across its systems, above and beyond the prevention of serious harms like incitement to violence and harassment? If so, who should make that decision? Should it be determined by an independent group of experts? Should governments set out what kinds of conversation citizens are allowed to participate in? Is there a way in which a deeply polarised society like the US could ever agree on what a healthy national conversation looks like? How do we account for the fact that the internet is borderless and that speech rules will need to accommodate a multiplicity of cultural perspectives? These are profound questions – and ones that shouldn't be left to technology companies to answer on their own. Promoting individual agency is the easy bit. Identifying content which is harmful and then keeping it off the internet is challenging, but doable. But agreeing on what constitutes the collective good has puzzled philosophers and politicians since time immemorial.

A crucial first step towards tackling these challenges is to acknowledge that they are not simply the fault of Big Tech's algorithms. Consider, for example, the presence of bad and polarising content on private messaging apps – iMessage, Signal, Telegram, WhatsApp – used by billions of people around the world. None of those apps deploy content or ranking algorithms. It's just humans talking to humans without any machine getting in the way. In many respects, it would be easier to blame everything on algorithms, but there are deeper and more complex societal forces at play. We need to look at ourselves in the mirror, and not wrap ourselves in the false comfort that we are the hapless victims of manipulative technologies.

As former President Obama acknowledged in a 2022 speech at Stanford University:

> [M]edia companies, tech companies, social media platforms did not create the divisions in our society, here or in other parts of the world. Social media did not create racism or white supremacist groups. It didn't create the kind of ethnonationalism that Putin's enraptured with. It didn't create sexism, class conflict, religious strife, greed, envy, all the deadly sins.

In the speech, Obama made a thoughtful case for technologists to put social responsibility alongside the profit-making motive, and to innovate to solve the challenges presented by the widespread use of these technologies. He was right. Both sides of this debate would benefit from a dose of humility. Technologies like social media and AI are not going to bring about the destruction of our democratic institutions, nor are they going to transport us all to some technological sunlit Elysium. The conversation we need to be having is the one Obama called for – a sober, reasoned and evidence-based dialogue about the impact of social media on society and the balancing of social responsibility and business imperatives.

The trouble is that sober dialogue feels a long way from where we are right now, and the reasons for that go very deep. Fundamentally, this is a debate taking place in a climate of fear: fear of the unknown, fear of change, fear of what we can't control, and fear of powerful entities that don't have our best interests at heart. But this is far from the first time a new form of technology has been met with fear and concern – in fact, most are. It's the real origin of these fears that we need to understand if we are to reach a place where such a dialogue becomes possible.

CHAPTER 2

The Techlash

'No one got upset when bicycles showed up. Right?' declares Tristan Harris in *The Social Dilemma*. 'If everyone's starting to go around on bicycles, no one said, "Oh my God, we've just ruined society!"' It was an interesting example for Mr Harris to choose. Because in fact that is exactly what did happen.

In the late nineteenth century, the arrival of the simple bicycle was greeted by a large number of people with nothing short of hysteria. A satirical 1894 *New York Times* article named 'Lunacy in England' skewered this view, claiming that 'there is not the slightest doubt that bicycle riding, if persisted in, leads to weakness of mind, general lunacy, and homicidal mania'. Riding bikes was blamed for skeletal conditions like 'bicycle foot' and 'bicycle hand', and even a condition called 'bicycle face' characterised by 'a hard, clenched jaw and bulging eyes'. It even changed the way women walked, distorting their gait into 'a plunging kind of motion'. Much of this hysteria focused on the bicycle's corrupting influence on women:

> The woman who rides a bicycle manifests her mental unsoundness by her dress. She may have dressed for years with decency and good taste, but after a few weeks' experience of the bicycle passion she is certain to manifest a desire to dress herself in the

most eccentric and conspicuous garments. The thirst for trousers which characterizes the advocate of woman suffrage is mild in comparison with the thirst for all manner of mannish and ugly garments which is developed in the feminine British mind by learning to ride the bicycle.

The *New York Times* article dramatised a genuine fear in the Victorian public. Yet there's nothing new about public backlashes against new technologies. In fact, it's hard to find an example of a new technology that didn't cause one. In the late nineteenth century, the telephone sparked fears of conditions like 'telephone ear' and 'telephone mania', of which 'an unmistakable symptom of the disease is a desire to talk to people at distant points about all sorts of things at all hours of the day and night', and it was condemned as a nuisance 'calculated to upset the nervous system of the most stolid and callous human being'. In the same century, the camera was seen by many as little more than a vehicle for spreading indecent and obscene images. Electricity in homes would make women and children vulnerable to predators who would know they were at home when the light was on. Reading novels was blamed for corrupting young minds and planting dangerous ideas in the minds of housewives. Hydro-electric power was a 'modern Frankenstein' and a 'suicidal folly'. Anti-vaxxers were rioting as early as 1881. Even the teddy bear was accused of 'destroying the instinct of motherhood in little girls', threatening to result in 'race suicide', the eugenics-tinged theory of the early twentieth century that a decline in birthrate among white American Protestants would cause them to dwindle and ultimately be replaced by other races (or, indeed, by Catholics). Later in the twentieth century, radio and movie crime dramas were said to be addictive to children and teens, who consumed them 'much as a chronic alcoholic does drink'. In the 1960s, violence on television was deemed a public health risk to children, much as violent video games were a generation later. As recently as 2014, British tabloids claimed that video games are 'as addictive as heroin', with the *Sun* declaring that 'Britain is in the grip of a gaming addiction which poses as big a health risk as alcohol and drug abuse'. Even the humble pager – the precursor to the mobile phone – provoked fears that young people were being given a direct line to drug dealers.

What all these moral panics have in common is a tendency to blame new technologies for endangering our health, corrupting our morality and diminishing our agency – particularly that of women and children. In other words, they are entwined with the fears and biases of those with power in society, whether it is men fearing female suffrage or white Protestants fearing they will become outnumbered. But, foolish as Tristan Harris's assertion about bicycles was, his underlying argument is that the reasons for today's backlash against social media are uniquely different and more profound than previous moral panics about technology. Is that true?

That question preoccupied me before my time at Facebook/ Meta and still does now. It would be comforting for technologists to believe that the techlash is simply history repeating itself – the latest example of a familiar cycle that will inevitably run its course. But while all breakthroughs in communication technologies have in some way enabled or speeded up the way we connect, interact and share news and information, the sheer scale and velocity of internet-based communications is unprecedented. The internet creates vast 'network effects' – the phenomenon whereby the more people are connected in a network, the more useful or valuable it becomes, leading to a feedback loop of explosive growth as more and more people connect. These effects have been accompanied by the new phenomenon of virality, in which content can proliferate exponentially as people interact with it and amplify it through their online networks of friends and followers, facilitated by algorithmic distribution systems. This phenomenon can spark social movements, grow businesses, turn ordinary people into global stars and help billions to share their joy, excitement and wonder as never before. But it can also ruin people's reputations, incite the anger of the online mob, and spread misinformation on an industrial scale, making the old adage that 'a lie can get halfway around the world before the truth has got its boots on' literally true in a way that Mark Twain – or any of the phrase's other purported authors – could never have anticipated.

Of course, in a sense, internet virality is a new form of a very old – and very human – behaviour. Rumours and gossip have always spread quickly through communities; they have always been carriers of half-truths, embellishments and outright lies, and they have always had the ability to enhance or destroy people's reputations. Some researchers have argued that gossip was one of the first forms of viral communication in early human societies, and was instrumental in creating an awareness of others and a shared sense of identity among community members. Even in those early communities, viral gossip could be either a manufactured or an organically emergent process; and it could be a tool either for social organising or for social shaming and ostracisation. So the dynamic is as old as the hills. But the network effects of a small village are pretty limited. The open internet, with its countless connections between the billions of people who use it around the world, is a different beast altogether.

Then there's the sheer ubiquity of data-driven technologies. We all have smartphones in our pockets. We spend hours every day looking at screens. Digital technologies have become embedded in every corner of our societies and our economies, to the extent that it is almost impossible for a citizen in the developed world to live a life untouched by the virtual world. And this has all happened in such a dizzyingly short space of time that we are only just starting to understand and grapple with what it means for our freedom, our careers, our children, our mental health, our rights as citizens and the functioning of our societies.

These factors pose big societal questions not only about the nature and role of these technologies, but also about the enormous commercial platforms that operate them. A small handful of companies have grown rapidly from tiny start-ups to global behemoths, making their investors and leaders supremely wealthy. With such far-reaching issues at stake, and such vast sums of money being generated, these internet platforms – and the people who run them – have become very consequential. Scepticism and suspicion are natural and healthy responses to concentrations of wealth and power. Scrutiny and accountability are unavoidable, even if it can be deeply uncomfortable for people once lauded as optimistic visionaries to find their every misstep exposed, and every decision prodded and probed under the microscope of public discourse.

No one will shed a tear for the Mark Zuckerbergs and Elon Musks of the world as the storms outside rattle the windows of their ivory towers. Nor should they. But one thing that's clear is that the backlash against technologies is particularly acute when they open up new ways to communicate and spread information. When Anne O'Hare McCormick, future winner of the Pulitzer Prize, questioned whether new technology explained 'all the furious fence-building, the fanned-up nationalisms, the angers and neuroses of our time', it was 1932 and she was writing about the radio. The same fundamental fear – that new technologies will stoke division and allow the spread of dangerous ideas – has greeted every major communications technology from the printing press to the internet. That's because there is nothing more controversial than human speech. Knowledge is power, and access to information and the ability to dictate public discourse are the essential currency of the powerful.

War of the words

Who does not know the story of the hysteria that resulted from Orson Welles's infamous radio broadcast of *War of the Worlds* in 1938? It was Halloween, and Welles directed and was the lead actor in a radio play, broadcast live on the CBS Radio Network across the United States, based on the H. G. Wells novel in which aliens invade earth. Only people didn't realise it was a play – they thought it was an emergency news broadcast. Mass panic ensued. Some people were reported to have committed suicide, such was their terror of the alien invasion. According to a 2013 PBS documentary, 'Upwards of a million people [were] convinced, if only briefly, that the United States was being laid waste by alien invaders.' The following day, the *New York Daily News* splashed an iconic front page declaring: 'Fake radio "war" stirs terror throughout US'. The *New York Times* howled: 'Radio listeners in panic, taking war drama as fact'. The *Boston Herald* cried: 'Thousands terrified by radio war drama'. The *San Francisco Chronicle* screamed: 'US terrorized by radio's "Men from Mars"'. A particularly alliterative *Washington Post* sub-editor called it: 'Monsters of Mars on a meteor stampede radiotic America'. It's a well-known story. But there's a problem with it. The panic didn't really happen.

CBS commissioned a nationwide survey the day after the broadcast. 'In the first place, most people didn't hear it,' CBS's Frank Stanton recalled later. 'But those who did hear it, looked at it as a prank and accepted it that way.' And it wasn't just the size of the audience that was exaggerated, but the panic itself. As Jefferson Pooley and Michael J. Socolow explain in *Slate* magazine:

> Wire service reports did relay sensational stories of (unnamed) panicked listeners saved only by the timely intervention of friends or neighbors, but not one newspaper reported a verified suicide connected to the broadcast. Researchers in Princeton's Office of Radio Research, working under the direction of Cantril, sought to verify a rumor that several people were treated for shock at St. Michael's Hospital in Newark, N.J. The rumor was checked and found to be inaccurate. When the same researchers surveyed six New York City hospitals six weeks after the broadcast, 'none of them had any record of any cases brought in specifically on account of the broadcast.'

The hysteria story was invented by America's newspapers. As historian and media critic W. Joseph Campbell wrote in 2010:

In short, the notion that the *War of the Worlds* program sent untold thousands of people into the streets in panic is a media-driven myth that offers a deceptive message about the power radio wielded over listeners in its early days and, more broadly, about the media's potential to sow fright, panic, and alarm.

Newspapers were accusing radio broadcasters of fake news – and in Welles's radio play they believed they had caught them red-handed. Pooley and Socolow again:

> How did the story of panicked listeners begin? Blame America's newspapers. Radio had siphoned off advertising revenue from print during the Depression, badly damaging the newspaper industry. So the papers seized the opportunity presented by Welles' program to discredit radio as a source of news. The newspaper industry sensationalized the panic to prove to advertisers, and regulators, that radio management was irresponsible and not to be trusted [. . .] Warned *Editor and Publisher*, the newspaper industry's trade journal, 'The nation as a whole continues to face the danger of incomplete, misunderstood news over a medium which has yet to prove [. . .] that it is competent to perform the news job.'

The fact that we all know – and almost all believe – the legend of the *War of the Worlds* panic shows just how effective the media can be at constructing and spreading a narrative. So much so that the legend is still repeated uncritically in textbooks, popular histories and documentaries like the PBS one. It is a remarkable demonstration of both the power of the press and the way the press responds when it feels its power is threatened by new technologies.

New technologies have always challenged existing power structures. And the powerful have never relinquished their power without a fight. So whose power is being challenged now? Just as the radio did nearly a century before, the internet has up-ended the business models of traditional news publishers.

According to the Pew Research Center, daily weekday news-paper circulation in the US was more or less flat from the 1960s to the early 90s, peaking at around 63 million in 1973 and again in 1984, before entering a sustained period of decline from the early 90s to the present day, falling to just 20.9 million in 2022.

From the nineteenth century onwards, news publishers have been the primary vehicle for politicians (both democratic and autocratic) to communicate to the public. This dependency has led, at times, to a toxic embrace between media magnates and politicians – something I've seen at first hand in UK politics. I remember, in the very early days of the coalition government, the *Daily Mail* editor Paul Dacre turning up in Downing Street. He met David Cameron and then came to my office. I hadn't met him before – the *Daily Mail* was not a fan of the Liberal Democrats – and I was advised that it would be a good thing for me to do so now. He then proceeded to lobby me – in private – against Rupert Murdoch's News Corporation's pro-posed takeover of the British broadcaster BSkyB. Here was the so-called tribune of middle England, who constantly published stories about the alleged corruption of Britain's political elite, lobbying me in secret against a commercial rival on his first – and, as it turned out, last – meeting with me in government. Dacre was later the Conservatives' favoured candidate to be the chair of Ofcom, the UK's media regulator, and was twice nominated for a peerage by Boris Johnson.

That's not to say I didn't play the game myself. Or try to. I have wined and dined newspaper editors and been to par-ties hosted by newspaper proprietors. But none of that stopped me being attacked by both the left-wing and right-wing press. Maybe I just wasn't very good at it.

As the power of the traditional media recedes, the certain-ties that politicians and media barons alike relied on have been thrown into doubt. And so the old media has an interest in discrediting the new forms of communications technology – the new business models that have up-ended their own – and in cast-ing suspicion on the newly powerful leaders of these companies who gain strength as their own drains away. The political classes too have an incentive to reinforce these narratives and prop up the old media that has served them so well, while simultaneously trying to figure out how to take advantage of the new media to rally support and spread their political message.

This hostility on the part of the old media goes beyond stoking their readers' and viewers' fears about tech. It's also about the bottom line. In Australia, Canada and elsewhere, traditional media publishers have exerted pressure on politicians to bring in laws requiring tech platforms like Meta and Google to subsidise them directly to the tune of tens of millions of dollars for the links to news content shared on the platforms' services, arguing that big internet platforms effectively steal original journalism for their own benefit, to the detriment of the traditional news publishers. The problem with this argument is that it fundamentally – and wilfully – misunderstands the value exchange between social media platforms and news publishers. When people share links to news stories on social media, they are directed to the webpages of the news publishers, which generates revenue for the publishers. There's no 'theft' involved – quite the reverse. Meta doesn't re-publish news stories on its own services: publishers make their stories available on their own websites and people are directed to them via social media. This is why publishers proactively share their stories on social media themselves and have buttons on their pages encouraging readers to share them. This is different, of course, from Google and other search engines, which have to scrape the internet for links to news stories, and snippets from them, in response to search queries.

In 2020, the year before Australia passed its News Media Bargaining Code into law, traffic from Facebook generated approximately 5.1 billion free referrals to Australian publishers, worth an estimated AU$407 million. Conversely, news links are a small part of the experience most users have on Facebook. On average, fewer than one post in every fifty in your feed will contain a link to a news story, and many users say they would like to see even less news and political content. These links are of little value to Meta, but of great value to news publishers. In other words, the attempts by Rupert Murdoch and other traditional media corporations to persuade politicians to force internet platforms to subsidise them amount to little more than an old-fashioned shakedown. But worse than that, as Tim Berners-Lee, the inventor of the World Wide Web, warned in the context of the Australian law, it 'risks breaching a fundamental principle of the web by requiring payment for linking between certain content online'.

Yes, the traditional news industry faces profound challenges. New technology has emerged, consumer behaviour has changed, and the ads-based business model of newspapers is increasingly anachronistic. But it makes no more sense to claim internet platforms are taking money from publishers than to say car companies stole from the horse and cart industry. Of course everyone wants quality journalism to thrive. Informed citizens are vital for healthy democracies. Whatever business models prove to be sustainable now and in the future, I firmly believe there will always be a market for high-quality journalism. This may prove to be even more emphatically the case in the age of generative AI when so much online content is synthetic, leading people to gravitate to trusted news brands.

A more acute problem comes a little lower down the food chain, in local journalism, which, broadly speaking, has been in decline for many years. Local newspapers and news sites provide an invaluable public service by holding local power brokers to account and sharing neighbourhood news, as well as providing a testing ground for aspiring journalists. If the old business model of local news doesn't work, how should it be paid for? Quality local journalism is a societal good, a societal necessity even, and when things are important for societies they need society-level solutions. In many countries, governments subsidise the arts because they recognise the value to society of a thriving cultural sector. Perhaps we should look at local journalism the same way. Governments could be actively looking for ways to support and promote local journalism, including subsidising it from central funds as they do other valuable public services. This would make far more sense than forcing one industry to subsidise another adjacent one – a precedent which undermines the principles of an open market economy so widely extolled, ironically, by the very publishers seeking a handout from Big Tech.

Techno-determinism

Regardless of how the economics of the news industry play out, stoking hysteria around tech creates demand for voices who will lend their support, instigating a cottage industry of polemicists in academia and the media. All of a sudden, critics of the new media find they are warmly welcomed and handsomely rewarded. As a result, previously nuanced critiques become ever more emphatic. Shades of grey become black-and-white clarity. Newspaper opinion pages are filled. Television talking-head slots are booked. Documentaries are commissioned. A narrative is cemented. The intangible fears people feel about the fast pace of change solidify as the culprits are identified and shamed. Virginia Tech science and technology professor Lee Vinsel writes about the phenomenon of critics and academic researchers becoming, in his words, the 'professional concern trolls of technoculture':

> The kinds of critics that I am talking about invert boosters' messages – they retain the picture of extraordinary change but focus instead on negative problems and risks. It's as if they take press releases from startups and cover them with hellscapes. At their most ridiculous, hype-filled criticisms become what historian David C. Brock calls 'wishful worries,' that is, 'problems that it would be nice to have, in contrast to the actual agonies of the

present' [...] But it's not just uncritical journalists and fringe writers who hype technologies in order to criticize them. Academic researchers have gotten in on the game.

In an essay for *The Atlantic*, Jonathan Haidt asserted, with fierce moral certitude and his own cherry-picked research, that the invention of Facebook's reshare button and Twitter's retweet (now renamed 'repost' in the era of X) is directly responsible for making us angrier, more divided and 'stupider' than previous generations. Not only is this strangely patronising, but it is asking us to adopt a radically deterministic view of technology as something with primal, god-like powers. In this school of thought, our anger is caused by the design of the communications medium through which that anger is expressed, and not by – for example – the economic devastation wrought on millions of people by the 2008 financial crash, the contraction of funding for public services that followed, rising economic inequality, systemic racism, sexism and homophobia in our societies, the huge stress and grief caused by the Covid-19 pandemic and the lockdown that accompanied it, the opioid epidemic, or the debt trap that millions of young people find themselves in.

The idea that the driving force behind people's choices and behaviour is technology, rather than the wider societal forces impacting their lives, is a common theme of backlashes against new communications technologies. This 'techno-determinism' implies that the machines are powerful and we are helpless. We are lambs to the slaughter, with no agency of our own. But technologies aren't inherently good or bad. They can, however, be used for good or bad purposes, by good or bad people. The printing press gave us the beauty of Shakespeare's sonnets and the hatred of *Mein Kampf*. The radio brought music into the living rooms of millions, and it gave dictators a captive audience for their propaganda.

Silicon Valley has itself to blame for legitimising the sort of techno-determinism now espoused by many of its critics. A tell-tale sign of techno-determinism is the desire to anthropomorphise technology, to ascribe to it agency and intent. Many tech companies insist on anthropomorphising their products by giving them human names, from Henry the Hoover to Alexa, from the voices on GPS satnav systems to the faces and 'personalities' of AI characters – not least Meta's own chatbots that look and sound like celebrities, including Tom Brady, Snoop Dogg and Kendall Jenner. Technologists have also been no strangers to making inflated claims about the impact of their technologies on the world. When the companies themselves tout a techno-determinist view of society, it's no wonder that people outside the tech world take this at face value. The two most unreliable groups of people to listen to about the power of technology are its fervent proponents and its fervent critics. They both exaggerate the deterministic role of technology on the human condition. A little more pragmatism on both sides would go a long way.

Many arguments made against social media today are based on a nostalgic desire to turn the clocks back. If only we could reverse technological progress a decade or more, the argument goes, we could preserve the internet in aspic at the point of supposed unity, before the fall of the Tower of Babel. Perhaps we could fix it at its 2011 peak, when the Arab Spring gave us reason to be optimistic about its potential as a force for good in society?

The connective tissue between techno-determinism and this backwards-looking nostalgia is good old-fashioned elitism. In this telling, people who express anger on Facebook and Twitter/X – complaining about identity-based discrimination, structural unfairness, or being 'left behind' economically and politically – have been induced to do so by the architecture of the platforms. But this version ignores the structural reasons for people's legitimate anger: inequality, poverty, racism, discrimination, and exclusion from power. Turning the clock back would mean reverting to a time when media was concentrated in the hands of a few gatekeepers who ensured the 'mirror held up to humanity' only reflected a culture dominated by one type of person: rich, white, male, western. This was an internet when only a few hundred million people were online, getting online was expensive, and smartphones were out of reach for many.

There's no recognition in this analysis that social media has given a voice to people who have been systematically overlooked over generations, and previously had no way of expressing themselves. You only have to look at the recent rise of the #MeToo and Black Lives Matter movements to see how social media has empowered previously unheard voices to speak out against deep-rooted injustice.

Turning the clock back to 2011, or to the late 1990s and early 2000s, before social media, would be convenient. Because if we went much further back, it would be hard to argue we were living in an era of peace and stability. Would the critique of social media as a defining factor in the polarisation of societies stand up to scrutiny in any other period? What about during the Vietnam War? The US was pretty divided then. Or the years of Jim Crow segregation? The McCarthy-era anti-communist witch hunts? Was British society less polarised during the Thatcher years, when miners were striking and people were rioting over the poll tax? It really does strain recency bias (the bias that gives greatest importance to the most recent event) to breaking point to suggest the levels of polarisation seen in the last ten years are somehow uniquely bad. In fact, it's the 1990s and early 2000s that have been the historical anomaly – in the West at least. You only need the most cursory understanding of history over the last century or so to see how bizarre it is to assert we are living in some kind of uniquely apocalyptic era.

What this critique really implies is nostalgia for a period of post-war prosperity in which middle-class, educated white men had a very comfortable time indeed – an era when there was consensus across elites, information was handed down paternalistically, largely by some of those same middle-class white men, sitting importantly behind desks in front of television cameras and watched by nuclear families sitting in cosy living rooms behind white picket fences.

America is not the world

Attitudes towards technology are often much more positive in the world outside the US and western Europe.

Regular surveys Meta conducts to gauge attitudes towards its brand and products show that people in Asia and Latin America are generally much more optimistic and positive towards the company. So, for example, when the company surveyed users in October 2024 asking if Facebook cared about them, the top six sentiment scores came from Thailand, Mexico, Brazil, the Philippines, India and South Korea. Users in these countries all scored Facebook in the top third of consumer brands, comfortably ahead of the UK rating and significantly higher than the US, which placed it in the bottom quarter.

When it comes to broader attitudes towards technology, Asia in particular is consistently more positive than the West. A large Oxford Internet Institute study (2020) on public perceptions of AI around the world found that just 25 per cent of respondents in Southeast Asia and 11 per cent of those in East Asia were concerned that AI would be harmful, compared with 47 per cent in North America. Another Pew Research study (also 2020) found broadly optimistic attitudes to AI among the Asian public, with about two thirds or more of people in Singapore, South Korea, India, Taiwan and Japan saying that AI has been a good thing for society. This optimism is often reflected in public policy. Citibank estimates that 'Asia adopts new technologies 8–12 years ahead of the West, making it much like a time machine to the future'. In the decade up to 2021, Asia accounted for 52 per cent of global growth in tech-company revenues, 43 per cent of start-up funding, 51 per cent of spending on research and development, and 87 per cent of patents filed.

These differences are evident when world leaders get together to discuss digital issues at summits. At the World Economic Forum's DX summit in April 2023, as European leaders warned about the apocalyptic risks posed to humanity by AI, Japanese digital minister Taro Kono brought his own digital doppelgänger – a full-size robot replica of himself like a Madame Tussauds waxwork come to life – to sit next to him. There was also a striking change in tone between the UK-led AI safety summit at Bletchley Park in 2023 and its remotely attended follow-up in Seoul the following year. While much of the debate at Bletchley – the birthplace of modern computing and wartime home of Alan Turing, the 'godfather of computer science', along with the team that cracked the Nazis' Enigma code – was dominated by concern about the potential existential risks posed by AI and the need to constrain AI development, the Korean summit focused more on AI's tangible, near-term risks and its potential benefits for national economies.

Why then are the United States and the West so glum about new technologies? Nirit Weiss-Blatt, a leading scholar on the techlash, traces the eruption of anti-tech sentiment to Donald Trump's first election victory in 2016:

> In the tech sector, there's a sentence that you hear a lot: 'change happens gradually then suddenly.' There were years and years of 'build-up' for the flip, but the flip itself was in the pivotal moment of Donald Trump's victory and the post-presidential election reckoning that followed it. The main discussion was the role of social media in helping him win the election. If Hillary Clinton had been elected in November 2016, the Techlash might have been much smaller.

The techlash represents a painfully US-centric world view – a howl of anguish from an American intellectual elite that is dismayed about what's happening within its borders. The fact is, the US has a wide range of deep-seated issues that are entwined with its history and its politics, and they contribute to the societal tensions it is experiencing today – from people on low and middle incomes feeling they've been left behind in the years since the financial crash, the lack of universal healthcare, and higher levels of labour-market insecurity compared with countries that have bigger welfare states, to its troubled history of race relations and its monied political culture that depends heavily on spending large amounts on negative campaigning. There are lots of reasons for Americans to feel angry and divided, and lots of reasons for educated middle-class white men to bemoan the loss of the trust and certainty that they've enjoyed for decades. But America is not the world, and what's happening in American society is not playing out identically everywhere else.

Disrupting power

One point on which many of the critics of social media are correct is that the individuals, institutions and systems that previously controlled who could speak and about what are weaker now. Whether social media is the cause, or an accelerant, or a beneficiary, or merely a historic coincidence, is far from clear. Undoubtedly social media turns traditional top-down control of information on its head, and far from being bad for society, I believe this is overwhelmingly positive. Disruptive? Yes. Uncomfortable for those in power? Yes. Better for those new voices not to be heard? No. Giving people tools to express themselves is a huge net benefit, because empowered individuals sustain vibrant societies.

The internet has also collapsed the gap between elites and the public. In the twentieth century, sources of information for the public were fairly few, so the traditional gatekeeper institutions in government, the media and elsewhere derived their authority in large part from their monopoly on public discourse. When access to information became democratised by the internet, those traditional elites lost control of the narrative. Their mistakes – which they always made, because they are human – became easier to expose, undermining their authority. And this genie won't go back in the bottle. It is a fundamental shift in the dynamics of power brought about by the internet, and, as author and former CIA media analyst Martin Gurri argues, we won't be able to make our peace with it until this generation of political and institutional leaders has handed over the keys to a new generation who are more comfortable in the hyper-connected internet age:

> Our politicians and institutions are going to have to adjust to the new world in which the public can't be walled off or controlled. Leaders can't stand at the top of pyramids anymore and talk down to people. The digital revolution flattened everything. We've got to accept that. I really do have hope that this will happen. The boomers who grew up in the old world and can't move beyond it are going to die out, and younger people are going to take their place. That will raise other questions and challenges, of course, but there will be a changing of the guard and we should welcome it.

The causes of people's grievances are deep, complex and real. Trust and social capital have been on the decline for decades in the US. Pointing the finger at social media, or the reshare button, may be comforting for those seeking an easy explanation for the legitimate fears they have about the divisions in America today, but it does little to advance our collective understanding of the real causes or what to do about them.

A common refrain among the critics of social media is that politics has been so fundamentally altered by virality that it is now impossible to reach agreement on anything. It is argued that we are so polarised by what we see online, so fearful of the Twitter mob and the so-called 'cancel culture' it inspires, and so misled by misinformation, that we can't renew the institutions of democracy. But this is what elites have said at every point in history when a new communications technology disrupts the status quo, or when power has been dispersed. But every time technological disruption has happened, institutions have adapted, sometimes painfully, to the new reality. There is no reason to think it will be any different this time.

And yet – and yet. Simply writing off all concerns about social media as moral panic and absolving tech companies of all blame is insufficient too. These are private corporations that have amassed enormous economic and social power. (We'll explore this in the next two chapters.) As an old-fashioned liberal, I have always believed that concentrations of power mustn't be allowed to go unchecked. In one form or another, these institutions need to be held properly accountable to the societies in which they operate. I'll examine how that might be done later on in this book.

Mixed in with the hyperbole of the techlash there are legitimate issues and concerns about Big Tech that will have to be addressed in some way. With billions of people using social media, all the good, bad and ugly of life is on display, and this scale brings with it a heavy responsibility for tech platforms to be mindful of their impact on society. The phenomenon of internet virality is new and represents a fundamental shift in how humans communicate. It is a consequence both of the network effects inherent in the internet, and of the way algorithmic systems make content available to the users of internet platforms. Tech companies can't shrug their shoulders and deny all responsibility.

Neither can they deny that while social media has democratised speech in a way that empowers people who have previously been denied the means to make their voices heard, there are also some people for whom the experience of being online has been damaging. Even if the latter are only a small proportion of all users, for platforms with millions or billions of users that will still be large numbers of real people experiencing real problems. The way social media usage interacts with a wide range of societal issues – politics, mental health, discrimination and much else – is complex and far from properly understood. Social media companies bear some responsibility for ensuring these issues are researched openly and independently, and for ensuring there are proper safeguards in place to prevent as much of this potential harm as possible.

Companies like Meta already do a lot in this way, some of which I'll describe in the coming chapters, but in tandem with policymakers and civil society they could undoubtedly do much more – and I'll explore this too. In the heat of the techlash, though, the popular instinct will be to rein in, curtail and cut down these global platforms, and remove their powers in order to reset their responsibilities. But let's not confuse today's imperfect practice with the underlying principles of the internet which led to these companies' power and size. We must not lose sight of the fact that virality is a function of something incredibly valuable. Before we take a scythe to Big Tech, we need to remind ourselves why it got big in the first place.

CHAPTER 3

The Open Internet:
How Big Tech Got Big

There is nothing inevitable about how technology develops. It is shaped by the ingenuity, tastes, values and preferences of the individuals who build it; the demands, expectations and behaviours of consumers; the commercial incentives of private companies; the strategic requirements of the state; and the regulatory environment in which these private companies operate. And these factors are themselves shaped by the values of the society and the politics of the moment in which the technological innovation occurs.

The internet is the most profound technological innovation of the late twentieth and early twenty-first centuries. It became so ubiquitous so quickly that it's easy to take it for granted, as if someone flicked a switch one day and suddenly we could all connect to everyone, everywhere, all at once. All the bells and whistles that came with it – from email and websites to apps, social media, e-commerce, streaming and video conferencing – can seem inevitable, as if their arrival were just a matter of bandwidth. But the global internet was not – is not – inevitable. It did not emerge fully formed as a single, interconnected whole. It was built, piece by piece, and joining these pieces together was a long, complicated process that required the development and adoption of compatible standards and protocols around the world. And embedded in these standards and protocols were principles and values.

There are different ways we can think about the internet. One way – the way we've been thinking about it in the previous two chapters – is as the public sphere. This is the lens we experience the internet through as users: a space in which competing world views collide; a space where political power – and the expression of identity – is constantly negotiated through language and code. Most of the highly politicised disputes about the internet today are about what can and can't be said, viewed or shared online and by whom, who is trying to influence whom and to whose benefit, and who gets to decide and according to whose rules. Through this lens, internet platforms are seen as not only hosts of the public sphere, but also participants in it when they engage in practices such as content moderation and algorithmic filtering.

But there is another, more fundamental way to look at the internet: as infrastructure. The internet is a collection of local and regional networks of submerged wires, data centres and platform software interconnected into a unified global system of communication. This infrastructure – the pipework of the open internet – ensures that an often cited but largely mysterious substance called data can flow freely all around the world, enabling all manner of interpersonal communication and economic activity. Seen through this lens, internet companies are understood as the providers of the pipework that enables the trade of ideas, goods and services both within and across borders. This may seem like a purely technical function. In fact, it is highly political too.

That is because this pipework was assembled largely through the ingenuity of Americans, with American expectations of free expression, free enterprise and freedom from government control. And those fundamentally liberal democratic values were embedded in the architecture of the internet. This 'network of networks' is designed to be 'net neutral' – to make no distinction between the types of information allowed to flow through its pipes. The global internet is an 'open' internet, and it is its openness that makes it the ultimate vehicle for free expression.

The nerds who invented the future

We have the internet of today because a number of people – a relatively small number in proportion to the impact they had – beavered away making things, argued about how they should work, had flashes of inspiration and moments of leadership, weighed up trade-offs and dilemmas, and ultimately created an architecture for what would become an interconnected network of networks that would provide the evolutionary starting point for the internet.

This was not a commercial venture. It was born out of a US government agency founded in the late 1950s by President Eisenhower: the Defense Advanced Research Projects Agency, or DARPA. DARPA's remit was to keep the US at the cutting edge of emerging technology. Spooked by the Soviet Union launching its Sputnik satellite in 1957, Eisenhower was determined not to be caught out again by America's tech-savvy enemies. But while the internet's creators were working for the military–industrial complex, they weren't themselves military people. They were scientists and mathematicians.

In August 1962, an MIT researcher called J. C. R. Licklider wrote a series of papers proposing what he called a 'galactic network' of interconnected computers. Two months later, he became the first head of DARPA's computer research programme. In 1965, his successor, Lawrence G. Roberts, working with another scientist called Thomas Merrill, connected a computer in Massachusetts to another in California via a regular telephone line. Buoyed up by the experiment's success, Roberts proposed a network of connected computers called ARPANET.

DARPA's military benefactors gave Roberts and his cohort of young scientists free rein. They brought to their work the mindset of freethinking academia – openness, cooperation, information sharing and problem solving – rather than the secrecy and hierarchy of military life. And this mindset was reflected in the network they created. While the strict, top-down world of the US armed forces and the loose, bottom-up world of academia seem like odd bedfellows, this counter-intuitive relationship was not entirely new, as Fred Turner notes in his seminal book *From Counterculture to Cyberculture*, about the relationship between counterculture revolutionaries and Silicon Valley technologists:

> [T]he same military-industrial research world that brought forth nuclear weapons – and computers – also gave rise to a free-wheeling, interdisciplinary, and highly entrepreneurial style of work. In the research laboratories of World War II and later, in the massive military engineering projects of the cold war, scientists, soldiers, technicians, and administrators broke down the invisible walls of bureaucracy and collaborated as never before. As they did, they embraced both computers and a new cybernetic rhetoric of systems and information. They began to imagine institutions as living organisms, social networks as webs of information, and the gathering and interpretation of information as keys to understanding not only the technical but also the natural and social worlds.

One of the young scientists working on ARPANET was Vinton Cerf, who came up with the Transmission Control Protocol (TCP) that would allow anyone to connect to a new network. He brought into the project his UCLA college friend and self-confessed 'math geek' Steve Crocker, who went on to lead the early internet oversight body the Network Working Group (NWG). This loose collection of grad students and other college researchers was crucial in setting the tone for the internet's culture. It wasn't to be a closed system accessible only to those with privileged access, but an open system that anyone with the right know-how and equipment could plug into. In her fascinating book *AI Needs You*, about the lessons AI can draw from the history of science, AI expert Verity Harding describes how this happened:

> It was left to this group to establish the philosophy of how ARPANET would work. It was a perfect example of the informality, decentralization, and collaborative approach that typified the internet's early development. The NWG made important technical decisions, including the foundations for email, but they also made social and cultural choices which have stayed with us as the internet has grown in size and importance, and which were inseparable from the values of those who made them.

Much of what is recognisable about the internet today flows from decisions and innovations made five decades or so ago by these often young and idealistic scientists. Thanks to them, the open internet transcends national boundaries, allowing data to flow freely between computer servers on every continent. It is built on the basis of common standards and protocols that have developed over time in that same spirit of openness and shared endeavour, making it work in the same way no matter where you are in the world.

For example, until the late 1980s each computer manufacturer would develop their own proprietary computer language for displaying images on a screen. This made sharing images between different computer brands technologically challenging. In the late 1980s, however, Steve Wilhite, a programmer who worked for a company called CompuServe, came up with a universal image-sharing computer language – GIF – that revolutionised the process of sharing images online. From then on, the images that users shared on CompuServe's network would be readable by any make of computer. Wilhite's innovation helped pave the way for the next generation of the internet, which was centred around the sharing of images, not just text.

Other internet technologies have longer and even more complicated histories than the GIF. Email, for example, has a history of more than fifty years of evolving technical standards. And the list goes on: sharing a video, creating a webpage, texting someone – each required a common technical language to be first developed and then adopted.

Today, the internet is open and accessible to billions of people because of pioneers like Licklider, Roberts, Cerf, Crocker, Wilhite and many others, and the work of international governance bodies like the Internet Corporation for Assigned Names and Numbers (ICANN), the Internet Engineering Task Force (IETF) and the World Wide Web Consortium (W3C). It is a particularly remarkable achievement when you consider that the role of governments in bodies like ICANN was deliberately circumscribed. Decision-making was explicitly divided up between governments, industry and civil society, with no single group having the power of veto. Neither industry nor civil society trusted governments not to try to dominate, and they successfully managed to restrain governmental power.

Assembling the pipework of the open internet in this way would never have been possible if it hadn't been built on open-source technologies. 'Open source' refers to software or code that is made publicly available for anyone to access, use, modify and distribute. Open-source projects are typically developed by a community of volunteers who contribute their time and expertise to create and improve the software. Anyone can view the source code of an open-source project, which allows developers to learn from it, identify and fix bugs, and add new features. Because the source code is publicly accessible, users can see exactly how the software works and check that it does what it claims to do, helping to build trust between developers and users.

This is fundamental to a large part of how the open internet operates. Open-source software powers servers, routers and devices that make up much of the infrastructure of the internet. Web browsers and apps built on open-source software, and written using open-source programming languages such as Python, shape how we as users experience the internet. The entire cybersecurity industry is built on open-source technology.

The alternative to open source is proprietary technology, which is developed, owned and operated by private companies or institutions rather than shared openly. Many of Apple's products, for example, are kept within its walled garden of hardware, iOS operating system and app store, which are inaccessible to people using PCs or Android devices. But while some companies and organisations may choose to keep elements of their technologies under proprietorial control – Meta included, although it has an institutional bias in favour of open sourcing where possible – the wider internet has open sourcing to thank for ensuring its component parts are compatible the world over.

This interoperability in turn makes the real secret sauce of the internet possible: its network effects.

Network effects and the rise of the platforms

The more people who connect to the internet around the world, the richer the experience. An internet with a handful of websites on it would not be much use and would get boring pretty quickly. But the more websites there are, the more information there is to search for and the more useful it becomes, and therefore the more attractive it is to spend time in. So new participants join the web, creating more websites and apps and so on. Given the low barriers to entry – all you need is a device and an internet connection – these new participants come thick and fast, and soon the internet becomes, among other things, a bustling marketplace of businesses, developers, content creators, platforms and consumers. The bigger the marketplace becomes, the greater the incentives are for developers to innovate for it, for existing businesses to grow their presence and experiment with ways to reach people, and for ambitious entrepreneurs and investors to pile in too.

These network effects explain the rapid spread of the internet as connectivity has improved and devices have become more affordable. And they explain the sheer scale of the digital economy. The whole thing becomes an enormous feedback loop, encouraging more and more people and businesses to shift more and more of their time and resources online, which speeds up innovation, enticing more people and businesses and so on.

These network effects also make big platforms more likely to emerge. With an ever-expanding amount of content and choice available online, the role of aggregators becomes essential. This is what the tech analyst and commentator Ben Thompson calls 'aggregation theory'. He argues that the value chain in a consumer market is divided in three – suppliers, distributors and consumers/users – and that in the pre-internet era many companies became big by integrating two of these parts. In many cases this meant merging supply and distribution. Newspapers created their content and distributed it to readers; book publishers controlled authors and distributed their writing; taxi companies owned fleets of cars and the ability to dispatch their drivers. But the internet turned this on its head:

> First, the Internet has made distribution (of digital goods) free, neutralizing the advantage that pre-Internet distributors leveraged to integrate with suppliers. Secondly, the Internet has made transaction costs zero, making it viable for a distributor to integrate forward with end users/consumers at scale. This has fundamentally changed the plane of competition: no longer do distributors compete based upon exclusive supplier relationships, with consumers/users an afterthought. Instead, suppliers can be commoditized leaving consumers/users as a first order priority. By extension, this means that the most important factor determining success is the user experience: the best distributors/aggregators/market-makers win by providing the best experience, which earns them the most consumers/users, which attracts the most suppliers, which enhances the user experience in a virtuous cycle.

This is one of the most significant ways in which the internet has disrupted power. As more and more of the consumer economy shifts online, the business models of companies that rely on owning the often expensive infrastructure of physical distribution are weakened. In a world where scale is the defining factor of success, rapid user growth takes precedence over monetisation, favouring nimble start-ups who spot a gap in the digital market for user-friendly services over slower incumbent businesses focused on the bottom line. Why would one take a time-consuming trip to a Blockbuster video store, with its expensively leased shopfloor and stockroom chock-full of costly physical inventory, when Netflix or Amazon Prime had vastly more movies available, which were never out of stock, and which you could stream directly to your television?

The same applies to social media. Users access it without charge, and both generate and distribute the content directly. By aggregating consumer demand, the platform can commoditise the supply of content created by its users. The better the social media platform is at attracting users, the greater the supply of content, which enriches the user experience, in turn making it more attractive to new users, who supply more content, which enriches the user experience, attracting more users and so forth. But these same dynamics also leave social media platforms vulnerable to competition. A new network that innovates in a way that attracts users – as TikTok did with short-form video sharing – can grow exceptionally fast. If the established players don't respond quickly, they can become irrelevant in no time – just ask MySpace and Friendster.

Social media is just one type of platform. The same aggregation theory explains the rise of big e-commerce sites like Amazon and eBay, streamers like Netflix, accommodation sites like Airbnb and Vrbo and ride-hailing apps like Lyft and Uber. In these cases, supply isn't free (as it is with user-generated content on social networks), but distribution is, making the aggregators that offer the best user experience the most attractive to consumers, which attracts suppliers, which in turn enhances the user experience – and so on.

This is why the borderless nature of the open internet is so crucial to its success. Borders restrict the user base, which limits incentives for suppliers, which slows innovation, which reduces the quality of the experience, which puts off consumers. The flywheel of network effects sputters.

The greatest symbol of globalisation

The internet is a product of the era in which it emerged. ARPANET only came about because of the US government's cold war-era technological rivalry with the Soviet Union. But it blossomed into the internet we know and love today thanks to the values of its nerdy pioneers, with their determined commitment to openness, collaboration and mass connection. It then reached critical mass in the late twentieth century, after the Berlin Wall had fallen in 1989. It both accelerated globalisation and benefited from it, providing a vehicle for greater economic and cultural integration. The world was getting closer together, in no small part because people all over the globe could find each other, communicate and share with each other, and buy from and sell to each other.

This gave rise to what we now call Big Tech. Nothing symbolises globalisation better than the big internet platforms. These platforms operate above and across national and regional jurisdictions, connecting citizens of different nations so they can communicate, share ideas and do business together according to liberal democratic (very much with a small 'l' and a small 'd') ideas of free expression and free enterprise.

There are differing definitions of what constitutes Big Tech. The most common is that it refers to the 'big five' technology companies: Alphabet, Apple, Amazon, Microsoft and Meta. Each had its own humble beginnings before at some point catching a wave and growing rapidly, and then expanding into numerous digital services. Alphabet owns Google, which started life as an internet search engine, but has since branched out into just about every corner of the online world – hardware,

operating systems, email, messaging, app stores, cloud computing, video sharing (YouTube) and other entertainment services.
Apple began life as a hardware company but has since branched
out into operating systems, email, messaging, app stores, entertainment, metaverse technologies and much more. Amazon
started out as an online retailer and is now the world's leading
cloud-computing supplier as well as providing entertainment
services, among other things. Microsoft started as a software
company and is now also in cloud computing, as well as email,
video conferencing, social media (LinkedIn), entertainment and
much else. Meta started as a social networking company before
branching out into messaging, video sharing, metaverse technologies and other areas. And all five are now very much in the
AI race. At the time of writing, these companies are together
valued at more than $9 trillion – around double the GDP of
Germany, the world's third largest economy. And in recent
years, the 'big five' have been supplemented by chip manufacturer Nvidia and Elon Musk's electric vehicle company Tesla to
become the 'magnificent seven'.

Big Tech is also often used as a broader term that captures
other large, albeit less diversified companies operating in the
same digital space – entertainment and social media companies
like Netflix, Twitter/X, TikTok and Snap. There is also a new
generation of AI companies like OpenAI and Anthropic that
has emerged too, capturing the public imagination and bringing
them into the conversation about the power of Big Tech. And
let's not forget that several traditional media companies have
also tried to transform themselves into online platforms – from
Rupert Murdoch's ill-fated purchase of the early social network
MySpace to Disney's more successful move into streaming with
Disney+.

While the digital economy has exploded globally, and governments have increasingly sought to bolster and grow their
domestic tech sectors, the primacy of Silicon Valley has until
recently largely remained untroubled. It is only in the last
decade that a number of Chinese competitors have stepped

onto the global stage as potential rivals, including e-commerce powerhouse Alibaba, multimedia giant Tencent and TikTok owner ByteDance. The last is the only non-US company given 'gatekeeper' status under the European Union's Digital Markets Act, which sets stringent rules on the biggest tech platforms, the significance of which we'll explore in a later chapter.

If you pause for a moment and think about it, the internet is not just a remarkable example of technological innovation. It's also a remarkable – and unprecedented – exercise in global, civic and industrial cooperation. It is the ultimate expression of, and exponent of, the liberal democratic values that characterised an era when barriers to free trade were removed, the rules governing the free movement of people were loosened, and the free exchange of ideas and culture was encouraged and celebrated.

But none of this was inevitable. Other significant twentieth-century scientific advances didn't evolve this way – often because of the political environment into which they emerged. Witness the differing ways in which in vitro fertilisation (IVF) was debated in 1970s and 80s America and the UK. In *AI Needs You*, Verity Harding describes how the highly politicised debate in the US over when life begins, influenced by the rise of the Christian right and the backlash to the *Roe* v. *Wade* abortion ruling in 1973, made IVF all but impossible for political leaders to regulate, forcing embryology research into the private sector and making access to IVF treatment both expensive for individuals and still politically controversial to this day. While a much less divisive issue in the US today – a broad consensus of Democrats and Republicans support it – IVF has not fully escaped the realm of politics, as demonstrated by the Alabama state Supreme Court's controversial ruling in 2024 that stored embryos should have the same legal protection as children, under a state law dating back to the nineteenth century.

In the UK, on the other hand, while IVF was highly controversial at first, the sting was largely taken out of the political debate by an independent commission established under Prime

Minister Margaret Thatcher – herself a trained chemist – and comprising scientists, doctors, lawyers, theologians, philosophers and non-experts, which established the framework for a regulatory regime that enabled both access to the treatment and innovation in the science. The UK was home to the first 'test tube baby', Louise Brown, born in 1978, and remains a scientific leader in this area today, thanks to the guardrails established as a result of serious parliamentary scrutiny and process. Perhaps more importantly, that scrutiny removed just about any sense of public controversy from the issue.

It's not as if the Christian right didn't exist in the UK in the 1970s and 80s, though – it certainly did, and Thatcher felt herself under considerable pressure from the right wing of her Conservative Party. But she also valued science and had a keen sense of which way the winds of public opinion were blowing. The idea of a royal commission into the subject came from the other side of the political divide – from the great former Labour education and science minister Shirley Williams, whom I was privileged to work with in her later years when she represented the Liberal Democrats in the House of Lords. It is to both women's great credit that they saw this as an issue above partisan politics. The fact it has never fully escaped political controversy in the US is a great pity. The different trajectories of IVF in the US and the UK show how much political pressures, processes and the decisions of individuals can distort technological progress.

And that has been the case with the open internet, which was shaped by the politics of the cold war and globalisation, the hands-off attitude of DARPA's military leadership and the liberal attitudes of a relatively small number of computer scientists. Having lived with it as part of our daily lives for three decades, it is easy now to take the open internet for granted. It is hard to imagine what the world would look like today without the ability to buy and sell online, to create and grow businesses without physical premises, to bank online, to stream music, television and movies, and much else besides. But we shouldn't take it for granted. It was shaped by politics and people, and politics and people will reshape it in the years ahead. Possibly dramatically.

Neither should we ignore the power paradox it creates. Economically, the internet's network effects, made possible by the creation and adoption of universal technical standards that allow data to flow freely across borders, have turbocharged a global digital economy that has transformed societies and industries, generating jobs, wealth and opportunities on an immeasurable scale in every corner of the globe. But the internet's inherent network effects have also aggregated hitherto unimaginable global reach in the hands of a few technology platforms. And the suspicion, discomfort and alarm that this power provokes is in large part what motivates and fuels today's techlash. So – what is the alternative? This is one of the most important questions we should be asking. And the answer is not hard to find.

A glimpse of an alternative history

China's attempts to control the internet would be like 'nailing Jell-O to a wall', declared President Bill Clinton in 2000. 'Liberty will spread by cell phone and phone modem,' he asserted; 'imagine how much it could change China.' There was just one problem. China figured out how to nail the Jell-O to the wall, and its internet model offers a glimpse at an alternative history.

For all intents and purposes, China's 'great firewall' now separates its citizens from the rest of the global internet. It imposes restrictions not only on content, but also on the flow of data in and out of the country – essentially creating a digital wall at its national border. Its pipes are still connected to the wider internet, but with heavy-duty valves that limit the flow to such a degree that very little can get in or out of the country unchecked.

China's internet model, like the wider open internet, was shaped by the politics of the moment – and it was far from the open, collaborative, laissez-faire politics of America's early internet pioneers. It dates back to the mid 1990s, just as dial-up connections were beginning to bring the internet into people's homes and workplaces. At this time, China was experiencing a population boom and was slowly opening itself up economically. But it was certainly not a democracy, and the ruling Chinese Communist Party (CCP) was concerned about outside influence on its culture, and on maintaining top-down control over its citizens. The party saw the potential to use computing to modernise the way its police forces gathered and organised information on Chinese citizens, but was also keenly aware that these fast-developing technologies could evolve more rapidly than its ability to control them. In 1996, despite only a tiny fraction of the population having access to it, a national order explicitly brought the internet under state control. China's Ministry of Public Security (MPS) created its Golden Shield Project, with the aim of integrating the files it kept on citizens into a nationwide digital registry and establishing a new form of data-driven surveillance.

As internet use grew in the early 2000s – from 2 per cent of China's population using the internet, to 77 per cent today – the MPS became increasingly concerned about the flow of information into the country. With the help of US tech company Cisco, it installed what are known as 'mirroring routes' at a series of choke points where fibre-optic cables entered the country, which made it possible to filter or block specific websites or keywords. This system evolved and became increasingly sophisticated over time, giving the Chinese authorities huge control over the content that is available to Chinese internet users. There are now more than sixty state agencies with the legal and technical ability to monitor and regulate online activity. And China's ambition to digitally monitor its citizens is no longer limited to what they can access on their internet browsers. As Alina Polyakova and Chris Meserole explain in a report for the Brookings Institution, in 2015 China's National Development and Reform Commission (NDRC) set the goal of establishing comprehensive CCTV coverage of all of China's public spaces and leading industries by 2020 in order to create an 'omnipresent, fully networked, always working and fully controllable' surveillance system. This led to the Sharp Eyes initiative (so called for the CCP slogan, 'the people have sharp eyes'), which aims to link smartphones, television sets and surveillance cameras. It has already produced apps people can use to monitor feeds on their phones and report suspicious activities. Polyakova and Meserole note that, by combining all of this with location data from phones and cars, 'Beijing will increasingly be able to monitor the movements and behavior of its citizens in unprecedented detail.'

The Chinese internet model not only means the state can clamp down on foreign influence and domestic political dissent, but it also keeps many US companies out of the Chinese internet economy. Many household-name internet services in the West, including Facebook and Instagram, are out of reach to Chinese internet users. 'International observers supposed that such control would stifle the ingenuity that led to the rise of IT hubs like Silicon Valley elsewhere in the world,' the security and emerging technology researcher Lorand Laskai notes. 'Instead, China's regime of online control has spurred its own form of domestic technological innovation and entrepreneurship, creating mini-Silicon Valleys across the country.'

No other nation got the same head start as China in creating the infrastructure for a siloed national internet while its user base was still small. But that hasn't stopped other countries, such as Turkey and Vietnam, from seeking to emulate elements of its approach. The most aggressive attempt to retrofit top-down state control onto the open internet has been in Russia, starting in earnest in 2014, long before the internet clampdown that accompanied its full-scale invasion of Ukraine in 2022. As Polyakova and Meserole explain, Russia's model could prove even more influential than China's:

Russian surveillance technology relies less on filtering information before it reaches citizens (as is the case in China) and more on a repressive legal regime coupled with tightening information control and intimidation of internet service providers (ISPs), telecom providers, private companies, and civil society groups. It is an ad hoc model utilizing legal, technical, and administrative means that is well-suited to diffusion across aspiring authoritarian regimes [. . .] In the long run, the Russian model may prove to be more adaptable globally as emerging authoritarian regimes that cannot afford China's high-tech model seek greater control over domestic populations and influence abroad.

The Chinese and Russian internet models should be cautionary tales for those of us in the West who take the freedoms of the open internet for granted. Cold war national interest, post-cold war globalisation and the openness and liberal culture of US academia combined to create the open internet. If the internet had first emerged instead in a political climate of deglobalisation and resurgent national sovereignty, it might have come to look more like the segregated, censored internets of China and Russia than a unified global network. And if that had been the case, the growth of the global digital economy would undoubtedly have been stifled, meaning that rather than an American-led global internet based on broadly democratic values, the rights of citizens online would vary wildly depending on culture and politics.

This is a glimpse of an alternative history, but also, as I will explore in later chapters, a possible future. And that possible future matters not only for the internet as we currently understand it. As I will discuss in Part Two, we are in the early stages of a new technological revolution that could be just as disruptive and transformative, if not more so, than the internet: that of generative AI. This new technology is emerging into a very different political environment from the one that nurtured the open internet, and there is no guarantee it will evolve in anything like the open and broadly democratic way the internet did. The implications of this are immensely significant.

In the meantime, however, the open internet presents the global technology platforms it has spawned with an unprecedented challenge. It is one that all of them – and Meta in particular – are now wrestling with.

CHAPTER 4

Nothing More Controversial Than Speech

Nothing is more controversial than speech, for hardly anything is more powerful. China's closed internet has been developed precisely in order to contain the challenge that speech represents. The open internet by contrast is premised on the belief that the freedom to express oneself in public is the lifeblood of democracy. In many societies that freedom is considered a fundamental human right. And, of course, the most radical democratisation of speech to have taken place since the birth of the internet – and one of the most consequential since the invention of the printing press – has been enabled by social media.

In 2012, the political sociologist Larry Diamond, of Stanford University, called social media a 'liberation technology'. Anyone with an internet connection and a phone can now express themselves, connect with people across geographical barriers, organise around common interests and share their experiences in an instant. The network effects of apps such as Facebook, Instagram, Twitter/X, TikTok and YouTube then make it possible for the right content, shared at the right time, to catch a wave and 'go viral'. Get it right, and something you post could be seen by millions, without an editor, network executive or news mogul's approval. Grassroots movements have grown rapidly to challenge established authority and orthodoxy – from the Arab

Spring to the Black Lives Matter movement and #MeToo – thanks to the ability these technologies have given ordinary people to share text, images and video in close to real time, and to have them amplified via networks of people connected through social media apps. Social media also made it possible for millions of spontaneous grassroots community-based groups to spring up during the Covid-19 pandemic to provide support for the vulnerable or to celebrate frontline health workers, as well as for millions of shops, restaurants and other small businesses to stay afloat and reach customers during lockdowns.

But let's not get carried away. The techno-utopianism of the Arab Spring phase of social media was never going to last. Social media interacts with politics and political issues in complex ways. Politicians use these tools to campaign, to rally support and to spread their messages. So do campaign groups, grassroots movements, private companies and all manner of other organisations across civil society and commerce. And so do ordinary people, expressing their political opinions and organising around causes they believe in. The result is often raucous and messy. When people make their voices heard, what some of them say can be deeply unpleasant. Racism and other forms of hate speech, bullying and harassment, and even incitement to violence, are all facts of online life. Likewise, people lie and try to manipulate each other, and when mixed with the viral dynamics of social media platforms, those lies can spread at a speed and scale previously unheard of.

These issues bring with them considerable dilemmas and trade-offs for both social media companies and governments. Who should decide what can and can't be said or shared online? Where should the line be drawn when people share unsavoury content? When should a private company remove content posted by an individual? Should private companies respect the rules set by authoritarian regimes who want to clamp down on dissent? And what should tech companies do when the President of the United States is the one posting offensive content on their apps?

Currently, in law, there are only very limited rules to guide internet companies' actions in trying to moderate content, especially for American companies rooted in a political culture born of the First Amendment's free speech protections. Protecting free speech was the intention behind one of the most significant early pieces of US internet lawmaking – Section 230 of the Communications Decency Act, passed in 1996. The law provides immunity to internet platforms from liability for content posted by their users, while allowing them to moderate and remove content they deem inappropriate. Without it, social media companies could be held directly responsible for any illegal or harmful content posted on their platforms by users. Given that these platforms have millions of users – billions in the case of Facebook and Instagram – posting content every day, the idea that the platforms themselves can make editorial decisions over whether every individual post should be published is clearly a fantasy. Without the protection afforded by Section 230, many of these services simply wouldn't be able to exist, and those that did would have to have much more restrictive and censorious content-moderation policies.

But this immunity comes at a price. Putting decisions over what content is and isn't acceptable into the hands of the platforms makes them a lightning rod for culture clashes between different views of what is socially acceptable. Different cultures and societies draw the line in very different places on what is acceptable public discourse. I've been berated, for example, by Scandinavian ministers aghast at what they see as American prudishness towards nudity. Conversely, America has an openness to gun culture that many in Europe and elsewhere baulk at. Holocaust denial has long been illegal in Germany, but is technically legal – if not socially acceptable – elsewhere. The greatest sensitivities around content in India concern ethnic identity and intercommunal violence.

Nor are these considerations static. Discussion about free speech and censorship has become much more fraught in the US as the Republicans and Democrats have become increasingly polarised on the issue, with the former suspicious of what they view as 'woke' censoriousness on behalf of Big Tech and the latter suspicious of Big Tech's everything-goes libertarianism. Both can't be right, or perhaps both are equally wrong, but the result is a nation deeply divided on one of the founding principles of an open society: the limits to free speech. In the UK and the European Union, on the other hand, content laws have been passed in the last few years that the First Amendment would make impossible in the US.

You can easily see why these differences cause so much heartburn for content moderators at social media companies. These are global platforms that want to treat their users equally, no matter where they are in the world, partly on principle and partly because applying different rules in different places is a devilishly complex task. But they also want their users to be able to express themselves in local contexts and be part of local online communities that operate within what are often very different idiomatic and cultural assumptions about what is and isn't acceptable.

In the great clash between technological globalisation and nation-based sovereignty, the idea that social media companies can operate with one-size-fits-all content rules was always going to be fraught. Even as Meta has sought to impose a consistent global rulebook – its Community Standards – it has had to make exceptions when certain forms of speech are illegal in local jurisdictions. That will increasingly happen as more and more nations write their own laws governing online content in the years ahead. It is worth looking closely at what this all means in practice, because the gap between what people want from social media and what the companies behind it can or do provide is at the very heart of the techlash. And the laws that get written in the heat of this moment have the potential to make the internet less open and less free.

The limits of free expression

If social media companies don't want harmful or hateful content on their services – a position that many people regard as illegitimate, for reasons we'll explore shortly – what, in theory, should they do about it? First, they have to define what those unacceptable forms of content are, and do so in a way that is reasonable and legally defensible within the expectations of the right of citizens to express themselves freely in the society in which they're based. For US companies like Meta, Snap and Twitter/X, that means American expectations of free speech, which, thanks to the First Amendment, are more absolutist than in most other democracies. These definitions of unacceptable content then also have to stand up to scrutiny when they are applied to a global platform with millions or billions of users outside the US (in Facebook's case, over 90 per cent of them are). That's not an easy line to walk.

Nonetheless, even within these limitations, there are some forms of harmful content that reasonable people everywhere can agree should not be permitted – child pornography, for example, or incitement to violence. Most would agree that content that glorifies terrorism should be taken down, though people can and do disagree over what constitutes terrorism. Similarly, most would agree that graphic violence shouldn't be allowed, but there are times when raising awareness of an atrocity or injustice can and should take precedence. And most would probably agree that adult pornography shouldn't be allowed either, although, as we've seen, attitudes to nudity more generally vary considerably in different cultures.

In my role at Meta, I oversaw, among other things, the teams that write and enforce the Community Standards for Facebook and Instagram. Whether Meta draws the line in the right place in its rules, or according to the right considerations, is a matter of legitimate public debate. But Meta certainly has a more extensive policy rulebook than Twitter/X, which in the Elon Musk era has lurched far further than Meta in the direction of unfettered free expression, resulting in more misinformation, hate speech and other forms of extreme content appearing on its services. It is also entirely reasonable to argue that private companies shouldn't be making so many big decisions on their own about what content is acceptable. It would clearly be better if these decisions were made according to frameworks agreed by democratically accountable lawmakers. But in the meantime, in the absence of such laws and frameworks, the decisions still need to be made.

In 2020, Facebook, as it was then, established an independent body – the Oversight Board – to make the final call on some of the most difficult content decisions on Facebook, Instagram and, later, its text-based social app Threads. The board is independent and its decisions are binding – they can't be overruled by Mark Zuckerberg or anyone else at the company. It is made up of experts and civic figures from around the world with a wide range of backgrounds and perspectives, including world leaders like former Danish Prime Minister Helle Thorning-Schmidt and journalists such as ex-*Guardian* editor Alan Rusbridger, alongside academics, lawyers and human rights experts. Acting like a sort of Supreme Court for content decisions, the board adjudicates on cases referred to it either by individual users or by Meta itself. As well as giving binding content decisions, it can also make non-binding policy recommendations when it thinks the company's broader policies are in the wrong place.

Having clear rules and enforcement processes is one thing. Live, high-stakes decision-making is another. No matter what rules you put in place, there will always be borderline cases where the right course of action is a matter of debate and judgement. These decisions are all the more loaded when they involve speech by influential public figures. And public figures don't get more influential than US President Donald Trump. Trump's posts have tested the boundaries of Meta's rules on a number of occasions, but two cases – both during Trump's first term in office – were particularly significant.

The first was at the end of May 2020, just weeks after much of America and the world locked down during the first major wave of the Covid-19 pandemic. Mass protests had sprung up in several big American cities in the wake of the killings of George Floyd and Breonna Taylor. On 25 May, George Floyd, a 46-year-old Black man, had been killed by Minneapolis police officer Derek Chauvin, who knelt on his neck until he suffocated – an appalling and grisly death that was caught on camera and spread like wildfire across social media. Breonna Taylor, a 26-year-old Black woman, had been shot in her home in Kentucky two months earlier by police officers. They were the latest in a long line of Black Americans to have been killed at the hands of police officers.

It's hard, several years after the event, to adequately express the emotions of the moment. Americans, like most of the world, had had their lives turned upside down for the last few weeks. This early period of the pandemic was filled with anxiety and grief on both a collective and intensely personal level, as well as boredom and introspection, often accompanied by economic hardship and uncertainty for those whose jobs or businesses had been either lost or put in jeopardy. There were no vaccines yet to offer protection, nor was there a clear understanding of the nature of the disease itself, or of how dangerous it was or how to prevent its spread. It was also an election year. President Trump had been in office for more than three years and the country felt as divided as it had been for generations. Culture war issues – from border walls and critical race theory to guns, abortion and trans rights – dominated the news day in and day out. And now these shocking deaths had provoked righteous outrage among much of American society. It was a combustible and highly politicised atmosphere.

Over several days, the protests in Minneapolis – where George Floyd lived and was murdered – and its surrounding areas grew increasingly volatile. Some remained peaceful, with people marching through the streets, blocking traffic and holding sit-ins. Others turned violent, with protestors smashing windows

and setting fire to businesses. There were clashes between protestors and police, with officers using batons, pepper spray and
other crowd-control measures to try to restore order. Into this
maelstrom, Donald Trump tweeted:

> I can't stand back & watch this happen to a great American City,
> Minneapolis. A total lack of leadership. Either the very weak
> Radical Left Mayor, Jacob Frey, get his act together and bring
> the City under control, or I will send in the National Guard &
> get the job done right.

> These THUGS are dishonoring the memory of George Floyd,
> and I won't let that happen. Just spoke to Governor Tim Walz
> and told him that the Military is with him all the way. Any dif-
> ficulty and we will assume control but, when the looting starts,
> the shooting starts. Thank you!

Many people understandably interpreted this as a threat of violence towards the protestors – or worse, a deliberate incitement
to violence against them. The phrase 'when the looting starts,
the shooting starts' has a specific history. In late 1967, amid
civil unrest in Black neighbourhoods in Miami, the city's white
police chief, Walter Headley, warned that his officers would use
shotguns and dogs to restore order, declaring: 'I've let the word
filter down that when the looting starts, the shooting starts.'

As he usually did, Trump reshared his tweets on Facebook.
Which meant the ball was in the company's court. Trump's post
was brought to the attention of its leadership by the company's
Strategic Response Policy team which is, as the name suggests,
responsible for ensuring the company responds quickly to controversial events or fast-escalating situations.

The decision had to be a judgement call. Essentially, there
were two ways Trump's post could be interpreted – one in
which Facebook's rules would require it to be taken down, and
one in which they wouldn't. If it were deemed a clear incitement
to violence, it would have to come down. On the other hand, it

could be read as a warning to the protestors that the state might shortly deploy force against them – a warning people clearly had a right to hear.

Removing a post by a sitting US president would not have been a small matter. Whatever your view of Trump, for a private company to choose to censor the democratically elected president would have been unprecedented and hugely controversial. So, was he encouraging violence, or warning people it was coming? Trump followed his original tweets with another insisting it was the latter – a warning. It was a finely balanced decision, and a series of conversations took place between various combinations of Mark Zuckerberg, Sheryl Sandberg, myself, Joel Kaplan, Facebook's VP of Global Policy, and Monika Bickert, the VP of Content Policy. I felt Trump had crossed a line and that the post should be taken down, but it wasn't an open-and-shut case. After a forthright debate, with people passionately arguing the case for each side, Mark Zuckerberg made the decision not to remove the post. He explained the decision in a Facebook post, saying:

> I've been struggling with how to respond to the President's tweets and posts all day. Personally, I have a visceral negative reaction to this kind of divisive and inflammatory rhetoric. This moment calls for unity and calmness, and we need empathy for the people and communities who are hurting. We need to come together as a country to pursue justice and break this cycle. But I'm responsible for reacting not just in my personal capacity but as the leader of an institution committed to free expression. I know many people are upset that we've left the President's posts up, but our position is that we should enable as much expression as possible unless it will cause imminent risk of specific harms or dangers spelled out in clear policies.

Facebook's policies were then tested again by President Trump a few months later, in the dying days of his first administration. On 6 January 2021, as lawmakers gathered in the US Capitol to ratify Joe Biden's victory in the previous November's

Presidential election, President Trump spoke at a rally of his supporters nearby. What happened next was nothing short of armed insurrection. Protestors stormed the Capitol building, fighting with police, breaking in and stalking the corridors and offices in a mix of menace and elation as Vice President Pence – whom some of the crowd had called to be hanged – and dozens of senators hid from the mob. In the feverish atmosphere leading up to the storming of the Capitol, Trump stoked the tensions in a series of tweets which he then reshared on Facebook, including one that said:

> These are the things and events that happen when a sacred land-slide election victory is so unceremoniously & viciously stripped away from great patriots who have been badly & unfairly treated for so long.

As events unfolded, and in the hours and days afterwards, Facebook's teams were on high alert for content that praised or encouraged violence at the Capitol or elsewhere. They searched for and removed content that praised or supported the insurrection, called for people to bring weapons to DC or to events and protests across the US, or called for violence elsewhere, as well as taking down videos and photos from the protestors. And while those of us in Facebook's leadership team were still wary of the implications of removing content posted by a sitting president, this time the case seemed clearer. Trump's support for those protesting at the Capitol, and his refusal to condemn the violence of the insurrectionists, was tantamount to inciting further violence, as well as a clear attempt to disrupt the peaceful handover of power. This time, Mark Zuckerberg made clear that the decision would be mine – as has been the case with other significant policy decisions over time. We suspended President Trump's access to his Facebook and Instagram accounts indefinitely.

The reaction to the decision showed the delicate balance private companies like Facebook were being asked to strike. Some said Facebook should have banned Trump long ago, and that the violence on the Capitol was itself a product of social media; others that it was an unacceptable display of unaccountable corporate power over political speech.

This was uncharted territory. Facebook didn't have a manual for what to do in the event that the President of the United States fomented a violent insurrection against his own government. I felt the indefinite suspension was an understandable action for the company to take at that time, but I was also acutely conscious that it was a big step for a private company, and one taken moreover without any precedent and without a clear process to follow.

This was a perfect case for the Oversight Board to rule on, and I referred the decision to it. After considering the case for several weeks, the board declared that it agreed with the decision to suspend President Trump's access to his accounts, but criticised the open-ended nature of the suspension, stating that 'it was not appropriate for Facebook to impose the indeterminate and standardless penalty of indefinite suspension'. The board instructed Facebook to review the decision and respond in a way that was clear and proportionate, as well as making a number of recommendations on how to improve its policies and processes. We did as instructed, establishing a new process to be applied in exceptional cases such as this, under which the maximum penalty would be a two-year suspension, not an indefinite one. This process was then applied to President Trump, including the full two-year penalty. In 2023, once the two years had elapsed, and with the risk to public safety that existed in 2021 having receded, the suspension was lifted. By this time, Trump had launched his own rival social network, Truth Social, and chose to do most of his posting there.

A few months before the Capitol insurrection, Facebook was facing a different set of content decisions that also had far-reaching consequences. As the Covid-19 pandemic unfolded around the world, people came to Facebook and Instagram to connect, commiserate, share opinions and look for answers. This was a global public health emergency, and with billions of people using Facebook's services all over the world the potential for spreading dangerous misinformation about the virus was clear. The company's leadership wanted Facebook's services to be a source of reliable information, but with events unfolding fast and precious little understood about the virus, that was not easy to achieve. So we did what we thought was common sense – looked to public health authorities like the World Health Organization and the US Centers for Disease Control and Prevention. In the early days of the outbreak in January 2020, the company began applying its harmful misinformation policy to claims which these experts told us were false and could lead to a risk of imminent physical harm (such as increased infection). Facebook then began removing posts containing a number of claims about masking, social distancing and the transmissibility of the virus. In late 2020, when the first vaccine became available, Facebook began removing claims about the vaccine being harmful or ineffective. By mid 2022, this policy provided for removal of eighty distinct claims about Covid-19 and vaccines. As a result, Meta – as the company was now known – removed Covid-19 misinformation on an unprecedented scale. Globally, more than 27 million pieces of content were removed between March 2020 and July 2022, more than a million of which were restored on appeal.

These policies proved to be controversial too. Not only did the expert advice change over time – for example, about whether or not wearing masks would be effective in keeping people safe or, later, about the origins of the virus in China – but as the pandemic became increasingly entangled with the wider culture wars, views on vaccines and mask-wearing became totemic for many on both sides. Meta was once again caught in the crosshairs, with some asserting that it was censoring the free speech of anti-vaxxers and other Covid-sceptics, while others denounced it for giving these voices a platform.

Content moderation has been controversial for years, but exactly what makes it controversial noticeably shifted in the time I was with Meta. Until a few years ago, by far the greatest public and political pressure applied to social networks like Facebook and Instagram in the US came broadly from the left. Our teams faced seemingly constant pressure to take more content down, to put more safeguards in place to prevent misinformation, and to do more, broadly speaking, to prevent divisive, angry or extreme content from going viral on Meta's platforms. This had particularly been the case since the election of Donald Trump in 2016, and the subsequent revelations about Russian attempts to spread disinformation through social media. The scale of this pressure represented what seemed to me quite a significant and illiberal shift in how a large segment of the organised left – particularly in the US – approaches matters of free speech. When I arrived in Silicon Valley, I perhaps naively assumed the First Amendment's free speech protections were a sacred cow for Americans across the political spectrum. What's more, standing up for people's right to say something, even if it is ugly, offensive or goes firmly against your own views, used to be a matter of principle for many on the progressive left. After all, it wasn't so long ago that the American Civil Liberties Union – long associated with progressive politics – had famously defended the rights of neo-Nazis to march through the village of Skokie in Illinois.

But in my first few years on the west coast it became clear to me that those attitudes had clearly shifted, to the point where many left-leaning US lawmakers, commentators and activists now felt that the only acceptable course for companies like Meta to take on bad – but still legal – content was to remove it completely.

A case in point is a video of then House Speaker Nancy Pelosi that went viral in 2019, which had been slowed down to give the impression she was slurring her words. It was misleading, mean-spirited, highly partisan and, especially to many Democrats, deeply offensive. But it didn't fall foul of any laws. Is this the sort of thing that should be excised completely from the public record in the land of the free? Meta has a number of levers to pull that can prevent problematic content from going viral without completely removing it. In this instance the company acted to significantly restrict the video's distribution across the platform – but not to remove it entirely. The decision not to take it down sparked outrage on the left and heavy criticism of the company.

It seems that every action has an equal and opposite reaction. In the late 2010s, and again in the aftermath of lockdowns and vaccine mandates in 2020, many on the right of US politics began exerting pressure in the opposite direction as they became increasingly energised about what they saw as left-leaning tech companies censoring conservative figures and causes. One man did more than any other to shift the Overton Window on content moderation: Elon Musk. When Musk set his sights on taking over Twitter, he did so with the explicit intention of cutting back on what he saw as overbearing censorship by his liberal tech peers. He used his newly acquired platform to rail against the 'woke mind virus' and either implicitly or explicitly to endorse a number of conspiracy theories and conspiracy theorists – from antisemitic tropes and climate scepticism to baseless statements implying Nancy Pelosi's husband Paul may have known the attacker who broke into their home or that British cave diver Vernon Unsworth, who saved the lives of twelve Thai children trapped in a cave in 2018, was a 'pedo guy'.

Of course, Musk's 'free speech absolutism' has proved to be somewhat less than absolute when it comes to people criticising him – for example when firing employees who criticised him or the company. The most valuable quality anyone can have when they put themselves in the spotlight of political debate is a thick skin. Musk's is clearly paper thin. But his impact on the political debate around content moderation – essentially normalising the idea that less content moderation is desirable – is undeniable.

For all the ugly consequences and glaring double standards that come from Elon Musk's version of free speech absolutism – not least the rampant hate speech, trolling and conspiracy theorising that has been amplified through Twitter/X since he took it over – there is clearly more than a grain of truth in his critique of social media. In a climate of public debate that had been dominated by criticism of social media companies for letting bad content go viral, they let the pendulum swing too far in the opposite direction, of censoring content, and in doing so opened themselves to criticism that they were political actors supporting a particular side's agenda.

The fact is that these unprecedented situations – the President of the USA posting offensive content and that which might potentially incite violence, and the spread of misinformation during a global public health crisis – brought into sharp relief some of the dilemmas social media companies face about how to moderate content on their platforms. Should private companies have the power to cut democratically elected political leaders off from services they use to communicate with the public, and if so, under what circumstances and according to what rules? Should social media companies be the adjudicators of what is true and accurate, and is it even possible to make these judgements when the facts are disputed? Should they make these decisions at all, or instead act simply as passive vessels for people to say and do what they like within the bounds of the law? Reasonable people can and do disagree over these questions. But decisions have to be made in real time, and in the absence of rules laid down in law, these decisions rest with the social media companies themselves, making them political entities whether they like it or not. And take it from me – they really *don't* like it.

War, dissent and free speech

The uses of social media – and the political choices facing the platforms – take on new meaning when authoritarian regimes either press it into service to manipulate public debate or clamp down on it to quash dissent. Both things occur in many regions around the world, and increasingly so during times of war and social unrest, when ordinary people use apps like Facebook, Instagram, Twitter/X, TikTok, YouTube, WhatsApp and Messenger to connect with each other within and across borders, to make their voices heard, to share news and information, and to organise and rally support. This has been especially apparent after Russia's full-scale invasion of Ukraine in 2022 and during the mass protests in Iran that followed the killing in the same year of Mahsa Amini, a 22-year-old Iranian woman accused of violating rules requiring her to wear a headscarf.

Within days of Russia's full-scale invasion of Ukraine in February 2022, Russian authorities attempted to block or restrict access to Facebook and Instagram in Russia as part of a wider attempt to cut their citizens off from the open internet, and to silence critics and independent media. State-controlled media outlets and Russian-based covert influence campaigns kicked into gear to spread propaganda and misinformation. In 2016, Meta (at that time still Facebook) had perhaps been slow to address such threats during the US Presidential election, instead focusing on more traditional cybersecurity threats like hacking, but since then the company has become increasingly sophisticated in how it identifies and takes down these campaigns, which often create vast networks of fake accounts to pump out misleading news and spread disruptive narratives with the aim of influencing elections, undermining one side or other of a political debate, or otherwise sowing confusion and distrust. Meanwhile other, less politically motivated campaigns are out to scam people. Since 2017, Meta has disrupted more than 200 of these so-called 'coordinated inauthentic behaviour' networks.

The widespread protests in Iran in 2022–23 prompted the authorities to clamp down aggressively on speech and freedom of assembly across the country. They also restricted the use of the internet and apps like Instagram. It's not hard to see why: Instagram was being used by millions of ordinary Iranians to spread word about the protests and the regime's brutal response. Less than six months after Mahsa Amini's death, hashtags related to the protests in Iran had been used on Instagram more than 160 million times. During the first three months of the protests, #Mahsaamini was the fifth-highest-trending hashtag in the world. People also shared Instagram footage of the protests with international news outlets, raising awareness of the events at a time when many media organisations couldn't report directly from the country.

These situations raise profound questions. Should private companies defend democratic values like free expression when they operate in undemocratic countries? This isn't a consequence-free philosophical question. Social media companies have employees based all over the world, and refusal to cooperate with authoritarian regimes can have real consequences for them and sometimes their families. Meta and other platforms receive a steady stream of requests from authorities all over the world – from democratic and authoritarian governments alike – to remove political content. These requests are often accompanied by threats of fines if the companies fail to comply, and are often justified with vague assertions of the need to maintain national security or public order. Meta frequently pushes back or refuses to comply with these requests if they are illegal or infringe on international principles protecting free expression. But refusal isn't without consequences. It can lead to these services being 'throttled', with internet connections slowed dramatically or blocked entirely. Some countries have proposed laws requiring internet companies to designate employees on the ground in those countries who can be held responsible by local law enforcement, which adds an unsettlingly personal element to any decision not to cooperate with government requests.

If resisting attempts by authorities to censor content on a company's apps comes at too high a price, the alternative is to withdraw services from that market altogether. Either way, however, free speech is restricted. Citizens are either limited in how they can express themselves on the app, or they can no longer express themselves on the app at all. These are tough choices for private companies to weigh up.

So content moderation is a delicate dance. Social media companies can perhaps be forgiven on occasions when they are perceived to be either too lax or overzealous in their approach to free speech, particularly during times of social upheaval and fraught public debate. But that is not to absolve them of responsibility either. There have been serious mistakes, not least during the genocidal campaign by the military in Myanmar against the Rohingya, which peaked in 2017. Hate speech and incitement to violence were spreading on its services, but Facebook didn't have enough Burmese-speaking content moderators and also faced challenges unique to the Burmese language that left it unable to keep up with the volume of dangerous content. (Burmese was largely written in a unique font-encoding standard, Zawgyi, which meant the standard Unicode-based tools used in the rest of the world would not work.)

While claims at the time that Facebook was completely unprepared for the situation didn't reflect the full picture – the company was energetically hiring Burmese content moderators, was working with local NGOs, and was seeking to share the Community Standards online and in Burmese tea shops, where people meet and socialise across the country – it had nonetheless failed to get to grips with the sheer quantity of disinformation and hate speech, and publicly acknowledged that it had been too slow to act.

A big internal culture change has taken place at Meta since then. With a more proactive, better-resourced and more culturally literate approach to Myanmar and other areas of conflict and heightened tension around the world, we can realistically hope that mistakes like this won't be repeated. But even if they aren't, social media companies can never escape the fact that wherever they draw the line between what content is and is not acceptable will always be deeply controversial.

And here again we encounter the power paradox. Not only have the internet's network effects both democratised and centralised economic power; they have also done so with power over speech. The open internet – and social media in particular – has enabled individuals to bypass the traditional gatekeepers of public discourse and make their voices heard in all their joyous, raucous, mad, bad and beautiful glory. Pandora's box has been opened, and all the virtues and vices of human life can be found online. But these democratising dynamics have also centralised power by putting consequential decisions over the ebb and flow of public discourse into the hands of those same democratising platforms. These private companies write and enforce their own rules dictating what can and can't be shared by the millions or billions of people using their platforms, but meanwhile are not directly or democratically accountable to those users. And unless a piece of content is specifically against a country's laws, the companies' power over speech isn't directly accountable to governments either, democratic or otherwise.

The inherent tension in technologies that simultaneously devolve power and concentrate it is the core reason why questions about how to fix social media or rein in Big Tech provoke so much angst. Something must be done to impose accountability on the powerful – but what? Would we accept a less free internet if it was a safer one? And if we did, how much freedom would we exchange for safety? Would we take back power from the ordinary users if that was the price of taking it from Big Tech? And if so, who would exercise that power in the future? In Part Four of this book I will set out how we might begin to tackle these dilemmas in a way that achieves the accountability we need while preserving the fundamental freedoms on which the open internet relies. This is a goal that I believe can only be achieved collectively, through international cooperation. But first we must turn our attention to a new and crucial piece of the puzzle that raises the stakes even further. For if these trade-offs seem difficult enough to resolve in the current age of social media, they are nothing compared with what is around the corner.

PART TWO

A New Power Emerges

CHAPTER 5

How Big a Deal Is Generative AI?

'Most of the time in politics, you spend time on stuff you think is important but it doesn't end up making a difference,' President Biden told me and other tech leaders privately at the White House AI summit in July 2023, adding, 'This AI thing is going to make the biggest difference.' About a year on from the release of ChatGPT, the then president had gathered senior figures from Amazon, Anthropic, Google, Inflection, Meta, Microsoft and OpenAI to thrash out a broad set of voluntary commitments to ensure AI is developed responsibly.

It's hard to overstate the shockwave OpenAI's ChatGPT sent through both the political world and the tech world. Silicon Valley companies and others had been developing AI technologies for quite some time. AI had been deeply embedded in Meta's content ranking, ads and integrity systems for years: the company had been releasing AI tools and models to researchers for nearly a decade, and generative AI was a core part of its strategy for building the immersive worlds and experiences needed to bring its vision for the metaverse to life. When Chat-GPT was released in November 2022, it was by no means the most powerful large language model (LLM) – the term for AI models that can respond to and generate text – available, but it was far and away the most user friendly, and it captured the imagination in a way nothing before had. Suddenly anyone could tap a few short instructions into a simple chat interface and experience AI for themselves.

The AI race that had been happening quietly in Silicon Valley suddenly roared onto the public stage, as Meta, Google and others raced to launch their own AI models and generative tools. For politicians, ChatGPT was the moment the scales fell from their eyes. Generative AI was a big deal, it was being developed at pace, and they needed to get on top of it. The near-universal mood among policymakers to 'do something' about AI was heightened all the more by the widespread view that politics had been late to get to grips with social media. Nobody wanted to make the same mistake again.

The 2023 White House summit – which resulted in tech companies signing up to a range of commitments around safety testing of AI models, prevention of bias and discrimination, privacy protections, and making it easy for users to tell whether audio and visual content has been altered or generated by AI – was preceded by a hearing organised by then Senate majority leader Chuck Schumer to kick off a process that he hoped would lead to bipartisan AI legislation. The UK government's own global summit on AI safety in November 2023 at Bletchley Park brought leading tech figures together with world leaders from the UK, US, EU, South Korea and elsewhere. A virtual summit hosted in Seoul took place the following year. The European Commission, never a slouch when it comes to penning new legislation, was already drafting its own AI Act and, after ChatGPT changed the terms of debate, sought to retrofit it with rules to address the issues raised by generative systems.

While some have greeted the recent breakthroughs in generative AI with dread about the risks these new technologies pose, there has been an equally predictable resurgence of giddy Silicon Valley utopianism too. AI, some believe, could help us cure cancer, harness nuclear fusion, solve the climate crisis or even live on Mars. Whether it will do any of these things remains to be seen. Even though I have been immersed in this world in recent years, I remain more of a pragmatic politician than a technologist. I believe it is probably best to stay away from the Kool-Aid and be as sceptical about the giddier claims as we should be about the predictions that AI is leading us to hell and damnation. I am, however, very optimistic that AI is going to accelerate scientific progress and make solving a lot of complicated problems significantly more efficient. AI will probably be the greatest problem-solving tool we've created so far – a pattern-recognition machine that absorbs vast quantities of information, analyses it, and creates something new from it in response to our instruction, and in doing so helps us to solve whatever challenge we've tasked it with.

We're already seeing AI-driven advances in science, research, healthcare and more. AI has been a catalyst for new advanced therapies and diagnoses for diseases like cancer and diabetes. It has helped reduce aviation carbon emissions, predict weather patterns, solve the protein folding problem, reveal the first ever images of unseen parts of the sun and decipher ancient scrolls.

AI could soon be an indispensable part of our relationship with our family doctors, as Natasha Loder, health editor of *The Economist*, notes: 'In the same way none of us would dream of getting onto a plane that didn't have an autopilot working alongside two human pilots, I predict there will come a time when getting a diagnosis from a doctor would be unthinkable without running it through a medical autopilot to check for human error.' Seeing the potential for AI to help us make better, earlier diagnoses and ever more scientific breakthroughs that could help us live longer, healthier, happier lives isn't utopianism; it's optimism grounded in the reality of how these tools are being used today and the likelihood of how they'll be used tomorrow.

But that couldn't be further from the view held by Tristan Harris and other techno-catastrophists. Never one to miss out on a bit of end-is-nigh hyperbole, Harris declared, 'What nukes are to the physical world [. . .] AI is to everything else.' At a

Senate hearing with tech leaders in September 2023, Harris said a team of engineers working on his behalf had used Meta's Llama 2 large language model to give instructions on how to make anthrax, only for Mark Zuckerberg to point out you can find instructions for that using Google search already. A little later in the hearing, Harris's colleague Aza Raskin sought to dispel the perception that they were anti-AI by enthusing about a project which is seeking to use AI to speak to whales. Whether the whales will welcome the intrusion is an open question.

As many commentators have pointed out, the idea of AI having the potential to create dystopian futures in which humans are subjugated or annihilated by robots has been a cultural trope for decades. But it has gained new traction in recent years, particularly since ChatGPT brought AI into mainstream conversation. Some believe so-called 'artificial general intelligence' (AGI) – AI more powerful than people, able to think, plan and execute actions free of human direction – is a fantasy that will never be achieved, the stuff of comic books and sci-fi movies and nothing more. Others believe AGI is both possible and desirable; that it will be benign by design; a more sophisticated tool capable of accelerating scientific progress beyond our wildest imaginations, but still, ultimately, a tool that humans will control. And some believe that a form of AGI super intelligence that outgrows humanity's ability to control it is at least theoretically possible. The truth is, technologists themselves can't agree on a definition of what AGI is and when we will know if it has been achieved, so the term becomes more unhelpful the more it is bandied about.

Apocalyptic claims about the existential threats posed by superpowered AI are speculative at best, and sometimes the people who make these claims have a clear self-interest in scaring the horses. A striking difference between the fears that have accompanied previous moral panics and the growing fear around AI is that with the latter it's often the proponents of the technologies themselves who promote the most apocalyptic scenarios. At various points, the likes of OpenAI's Sam Altman, Google DeepMind's Demis Hassabis, Elon Musk and others have all talked up the dystopian future their own technologies are contributing to. An open letter from the Musk-backed Future of Life Institute in 2023 called for a six-month 'pause' on all AI development because of the existential risk the technology poses. But despite the roll call of luminaries who signed it, the letter lacked any clear argument or any evidence that this risk exists, offering just 'a lot of highly speculative assumptions', according to the techlash scholar Nirit Weiss-Blatt in an article entitled 'The AI doomers' playbook'.

Altman has talked up fears of authoritarian regimes using AI for cyberattacks, and of AI turning on humans and attacking us. He has even gone so far as to stockpile his own survival gear in true Silicon Valley style: 'I try not to think about it too much [. . .] But I have guns, gold, potassium iodide, antibiotics, batteries, water, gas masks from the Israeli Defense Force, and a big patch of land in Big Sur I can fly to.' So why all the doomsaying? Ben Thompson has a theory:

> The point is this: if you accept the premise that regulation locks in incumbents, then it sure is notable that the early AI winners seem the most invested in generating alarm in Washington, DC about AI. This despite the fact that their concern is apparently not sufficiently high to, you know, stop their work. No, they are the responsible ones, the ones who care enough to call for regulation; all the better if concerns about imagined harms kneecap inevitable competitors.

Even Altman's old professor at Stanford, Andrew Ng, has expressed scepticism about the motives of AI doomsaying by tech leaders, telling the *Australian Financial Review* that the 'bad idea that AI could make us go extinct' was merging with the 'bad idea that a good way to make AI safer is to impose burdensome licensing requirements' on the AI industry. He went on to say, 'There are definitely large tech companies that would rather not have to try to compete with open source, so they're creating fear of AI leading to human extinction. It's been a weapon for lobbyists to argue for legislation that would be very damaging to the open-source community.'

Commercial self-interest, hyperbole and paternalism are a heady mix. In the months after the launch of ChatGPT you could hardly turn on the television or scroll your phone without being greeted by wild apocalyptic claims about AI. It's telling, however, that the doom-mongering died down almost as fast as it arrived. It's hard to imagine Musk's open letter getting half as many signatures today. All the hype and activity that followed ChatGPT's arrival were almost certainly amplifying the most alarmist voices. I made this point during the Bletchley Park summit to Vice President Kamala Harris, British Prime Minister Rishi Sunak, European Commission President Ursula von der Leyen and others. I asked them to imagine an entirely hypothetical Mrs Miggins, living at 63 Orchard Close down the road in Milton Keynes. 'I can guarantee,' I said, 'that she's more terrified of AI now than she was before this summit started two days ago. How is that going to help public acceptance of AI? It's not clear what we have achieved by hyperventilating about existential risks and scaring the living daylights out of people.' In his closing remarks, Rishi Sunak graciously echoed the sentiment, saying, 'To quote one of my predecessors, I agree with Nick,' a reference to Gordon Brown in the 2010 televised UK party leaders' election debate, when he agreed with me so often that the phrase 'I agree with Nick' became an internet meme (an all-too-fleeting high point in my political career).

I also made the point at Bletchley Park that it's important to distinguish between today's AI models and the potential models of the future. The most dystopian warnings about AI are really about a technological leap – or several leaps – beyond where we are now. There's a world of difference between the chatbot-style applications of today's large language and multi-modal models and the supersized frontier models theoretically capable of sci-fi-style superintelligence. Terrifying as the apocalyptic warnings are, we're merely in the foothills, debating the perils we might find at the mountaintop. If and when these future advances become more plausible, they may necessitate a different response. But we are still some way off that point, if we ever reach it at all.

More urgent is the more prosaic work of mitigating the risks associated with today's AI technologies – not least the potential for deepfakes and AI-generated misinformation to mislead people, the potential for criminals and foreign governments to use it to hack through cyberdefences, and the ability of AI systems to inherit biases that exist in the data sets their models are trained on. In the next chapter, I address the broader question of whether and to what extent the risks are balanced or outweighed by the opportunities and benefits that may be made available by generative AI, but the key point here is that while these are challenges for the tech companies themselves, they are common problems across the wider industry which require common solutions.

In a world where so much content is synthetic, I think it is inevitable that in time most people will develop a healthy scepticism about what they encounter online, much as we have become more likely in recent years to assume that images of models on magazine covers have been airbrushed or Photoshopped. But healthy scepticism alone won't be enough to prevent generative AI assistants from being used to spread persuasive personalised misinformation, or to scam and defraud people. As the *New York Times* points out, 'personalized, real-time chatbots could share conspiracy theories in increasingly credible and persuasive ways [by] smoothing out human errors like poor syntax and mistranslations and advancing beyond easily discoverable copy-paste jobs'. What's more, generative AI can become exponentially more harmful for people who have already developed a trusting relationship – and possibly an emotional attachment – to their AI assistant. Policymakers and tech leaders are going to have to be clear-eyed about the challenges we face and the safeguards that are necessary to prevent or mitigate the worst abuses.

That makes collaboration across industry, government, academia and civil society essential. For example, Meta is a founding member of the cross-industry Partnership on AI, and is participating in its Framework for Collective Action on Synthetic Media, an important step in ensuring detection and provenance guardrails are established around AI-generated content.

Governments have a big role to play too. It's encouraging that many countries are considering their own frameworks for ensuring AI is developed and deployed responsibly (the White House's 2023 voluntary commitments, for example), but it is vital that governments, especially democracies, work together to set common AI standards and governance models.

Generative AI has the potential to be both sword and shield for digital platforms. It can help bad guys spread problematic content, but it also helps companies and others get better at identifying and stopping the spread of this content. Meta deploys AI systems aggressively to reduce the prevalence of hate speech and other problematic content on its platforms. The company's years of experience developing AI models and tools have helped it build safeguards into its products from the beginning, for example by training and fine-tuning models to fit its safety and responsibility guidelines. And, crucially, the company's highest-profile models are thoroughly stress-tested by conducting what is known as 'red teaming' with external experts and internal teams to identify vulnerabilities at the foundation layer and help mitigate them in a transparent way.

Then there is the problematic question of energy. Today's AI systems are resource intensive. They require vast data centres, which are hugely expensive to build and operate, take up large plots of land, and require enormous amounts of water to keep them from overheating and a lot of energy to run. A recent study by Hugging Face (a company that, among other things, hosts leaderboards to track the performance of AI models) and Carnegie Mellon University estimated that generating a single

AI image uses as much power as fully charging an iPhone. Electricity consumption at US data centres alone is poised to triple from 2022 levels by the end of the decade. That's the equivalent of about 7.5 per cent of the projected electricity demand for the entire United States.

While all computing systems require energy, it is particularly critical for AI systems. For a long time, the huge data centres required for cloud computing were becoming increasingly efficient, even as demand was rising significantly. From 2010 to 2018, global computing output in data centres jumped six-fold, but their energy consumption rose by only 6 per cent. But the graphics processing unit (GPU) chips required for LLMs consume much more energy than the central processing units (CPUs) that have traditionally powered data centres.

Energy consumption on this scale must not be ignored by policymakers. The climate crisis is real and needs urgent collective action. We need to be reducing our carbon energy consumption, not ramping it up. This is just one of the many reasons, which I will explore in later chapters, why the race for AI capacity is now firmly a matter of geopolitics and not simply of private enterprise.

All of this is predicated on the idea that the AI race will continue to be focused on building ever bigger and ever more powerful AI models – and that energy inefficiency will remain where it is today. However, there is a school of thought that suggests this might not be the case. If past technological development is any guide, significant incremental improvements in the energy efficiency of these technologies are likely. Already, chip manufacturer Snapdragon and Google have been pushing to incorporate some parts of the computational power required to run AI into mobile phones and away from data centres.

There is also the prospect that LLMs could hit a developmental ceiling, reaching a stage beyond which improvements are marginal, making further development too costly to be worthwhile. OpenAI's GPT-4 took around $100 million to train, but it has been estimated that the next generation of models could cost a billion dollars each. And then there's the problem of the diminishing amount of new data available for model training. Many models have already absorbed the publicly available text on the internet, and there's a limit to how much exists in other forms – licensed content, video transcripts, and so on – to add to the pile. This has led many AI companies to generate their own synthetic data: new data generated by AI to fill in the gaps, like the frog DNA used by the *Jurassic Park* scientists to bring dinosaurs back to life. But it's hard to know how sustainable this data inbreeding is, and whether it will start amplifying defects like a digital version of the Habsburg jaw.

One argument, most consistently articulated by Yann LeCun, Meta's chief AI scientist and a winner of computer science's highest accolade, the Turing Award, is that the future path of AI development is to design models that mimic the way human and animal brains learn – in essence, building a mental model of how the world works by extrapolating from the things we observe and experience – and therefore operate far more efficiently, digesting less data, than today's LLMs. According to LeCun, these models will also be easier to constrain, and be less inclined to 'hallucinate' by fabricating wrong answers, which would help allay fears about their impact. LeCun has provided an early glimpse of what the future of AI might be through the development of Meta's I-JEPA AI model. His approach promises an AI that is less reliant on large training data sets, requires far less energy and is less inclined to make things up. I for one certainly find LeCun's outlook reassuring. But it is not uncontested among his peers, and at this stage we have no way of knowing whether this alternative approach will win out.

In my view, it is these perhaps more prosaic concerns – bias, security, misinformation, energy – that we should be focused on right now, rather than eye-catching assertions of apocalypse and dystopia. They are, at least in principle, risks that can be mitigated, if not solved entirely. That is not to say that mitigating them will be easy, and, when it comes to the use of AI for nefarious purposes, it will inevitably involve an iterative process as governments, tech companies and law enforcement agencies evolve their tactics and hone the technology to try and stay one step ahead of their adversaries. Those concerns need to be addressed by tech companies, governments and civil society collaborating to develop common standards and safeguards on the one hand, and by regulation on the other. Unglamorous work of this kind isn't likely to excite the headline writers, but it is both necessary and possible. And it begins with an understanding of what living with these technologies in the very near future will actually look like.

Living with machines

Today's generative artificial intelligence models aren't 'intelligent' in the way that you and I would understand. They have no inner life, no independent agency. They are purely responsive. In so far as they 'think', they do so in a probabilistic way. They make predictions. They are, in effect, highly sophisticated pattern-recognition machines – advanced autofill programmes capable of guessing the next link in a chain based on prompts given to them by people. This is what a large language model does – it analyses millions, if not billions, of data points and spits out a response. These 'foundation models' can then be trained and customised to perform specific tasks, which they often do with a remarkable speed and efficiency. Generative AI systems also 'learn' over time, fine-tuning their responses as they gain greater familiarity with the context of the prompts they are given. But they're not conscious. They're not thinking for themselves like we do.

The form of generative AI that you and I are most likely to interact with over the next few years is what is referred to as an 'agent'. Agents are personalised AI assistants that live in your phone or your computer (or in your smart glasses), that will not just respond to your questions but will also carry out all manner of administrative tasks on your behalf: drafting emails, creating documents, synthesising and summarising information. Their uses will go well beyond the workplace. Today's AI chatbots are relatively primitive. Most are effectively single use, as they live in a single app and respond only when you prompt them to do so. And most have no memory of previous tasks and interactions with you, while those that do learn from previous interactions only do so in a limited way. Based on what permissions you give them, agents, or assistants, will be able to operate across a range of apps, accomplishing a much wider range of tasks. They will be able to remember your interactions, learning your routines, interests and inclinations in a much more sophisticated way. As a result, they will not only make recommendations for you but also execute tasks on your behalf. As Bill Gates describes:

> Whether you work in an office or not, your agent will be able to help you in the same way that personal assistants support executives today. If your friend just had surgery, your agent will offer to send flowers and be able to order them for you. If you tell it you'd like to catch up with your old college roommate, it will work with their agent to find a time to get together, and just before you arrive, it will remind you that their oldest child just started college at the local university [. . .] Agents won't simply make recommendations; they'll help you act on them. If you want to buy a camera, you'll have your agent read all the reviews for you, summarize them, make a recommendation, and place an order for it once you've made a decision. If you tell your agent that you want to watch *Star Wars*, it will know whether you're subscribed to the right streaming service, and if you aren't, it will offer to sign you up.

Imagine a day in the life of a working couple. Let's call them Clarice and Matteo. They have two school-age kids and demanding jobs. Like all working parents, they will be able to use AI to simplify a range of tasks throughout their day, especially those that already require using some kind of technology.

The couple wake up to the sound of a new song that Clarice has loved ever since her personal AI assistant recommended it to her two days ago. In the kitchen, Matteo asks his assistant for a quick and easy breakfast recipe he can prepare for Clarice and their kids, Laura and Pietro, based on what they have in their refrigerator. Meanwhile Clarice asks her assistant to help her prepare for her commute by checking the traffic and weather situation and proposing the best route to school, then on to work and, later in the afternoon, to a doctor's appointment.

At work, Clarice's AI assistant helps her create first drafts of emails, slide decks and other documents, schedule meetings and other appointments, and prompts her on the optimum way to use gaps between meetings to clear the tasks from her to-do list so she can focus on the more creative aspects of her job. Matteo prefers to do as much as possible without the help of an AI assistant, but secretly assigns it a handful of administrative jobs to free up some time.

At the doctor's, Clarice learns that new AI advances helped her doctor make an early assessment of her chances of developing cancer later in life. They are able to discuss some cancer prevention options, and her doctor also gives her tips on nutrition, which she instructs her assistant to incorporate into her weekly shop. She also instructs her assistant to pass the nutrition tips on to Matteo's assistant.

Later, at the dinner table, Matteo asks Pietro about his homework and speaks to his personal AI assistant to get some ideas on how to help his son with history class, which he inexplicably seems to find boring. Once the kids are in bed, Clarice asks her assistant to learn more about healthy illness-preventing nutrition. She also directs it to recommend the best social media support groups on this issue. She asks her assistant to draft a message to post to the community it has picked, since she feels uncertain of what exactly to say. Soon after she posts her message, she receives a friend request from someone in the group, and several people 'heart' her message, leaving comments of support. After Clarice nods off, Matteo, sick of the song Clarice has chosen for the morning alarm, instructs his assistant to liaise with hers and pick a new song based on their joint preferences. It is not clear whether his assistant will manage that – AI can solve a lot of things, but differences in musical taste between partners may not be one of them.

None of these tasks are things Clarice and Matteo would have been unable to do by themselves, and many of them would still involve online interactions if they did. Some of them are things others might prefer or even enjoy doing without any assistance: we don't all want help finding music or writing messages. But we all have things on our to-do list we never quite get to, and pesky but necessary tasks that clutter our days. The couple's assistants take the clutter off their plates, just as a personal assistant frees up an executive to focus on their most important work. And, of course, the net effect of removing these time-consuming tasks is not just added convenience and less stress: it should also allow people to focus on the things that matter most to them, whether spending time with each other and their children, or pursuing interests and hobbies, or any number of other things that busy people would do if only they had the time and headspace to do them.

While many of the day-to-day examples I've just described are discrete tasks, AI assistants will also be able to carry out multi-step projects on our behalf. For example, Clarice could instruct an AI assistant to search and give recommendations for a holiday in Greece from start to finish – flights, hotels, restaurants, attractions and local transport. Of course, she would expect the AI assistant to run its proposed plan past her first, but with her approval it would have permission to then execute a chain of actions for her, including negotiating with other people (or AIs), making payments and even adjusting the plan on the fly, just as a good human personal assistant would do.

Examples like these are realistic in the immediate future. In the longer term, it is conceivable that AI assistants will be able to do much more for us, and potentially without the need for us to prompt them to do so. If, for example, we gave them permission to access cameras at home, or in our smart glasses, or to sync with cameras in the workplace, then it is possible that they could anticipate our needs based on the context they observe and take actions proactively on our behalf. This sort of relationship with AI assistants is not without its own policy quandaries, particularly around privacy and liability for actions taken for us by machines, and these are issues society (and lawmakers) will need to grapple with sooner than they might imagine.

AI assistants can also operate as personalised co-pilots that support us during difficult periods in our lives. One of the pioneers in the field of 'affective computing', MIT professor Rosalind Picard, has recently begun testing ways to utilise AI assistants for mental health. She argues that generative AI can be useful to individuals suffering from depression and anxiety. As she notes:

> Since there aren't enough therapists for people suffering from depression and anxiety – the two largest sources of mental illness – we're looking at what can be done with technology to support those people. And in particular, we're thinking about not just healthcare, which is misnamed because it's just really sick care, but about real healthcare – before people are depressed, diagnosed with Major Depressive Disorder or anxiety disorder.

In an article for NPR, consumer health correspondent Yuki Noguchi tells the story of Chukurah Ali, a baker who owned her own small business – Coco Desserts in St Louis, Missouri – making bespoke wedding cakes. Her life was turned upside down when a car accident left her seriously injured and unable to work. Without health insurance, Chukurah couldn't afford therapy for the depression that ensued. Her physiotherapist recommended a mental health app called Wysa, which its founder describes as a friendly and empathetic tool that asks people simple questions – 'How are you?', 'What's bothering you?' – and gives answers from a database of messages pre-written by a qualified psychologist.

> Ali says that as odd as it might sound to some people, after nearly a year, she still relies on her chatbot. 'I think the most I talked to that bot was like seven times a day,' she says, laughing. She says that rather than replacing her human health care providers,

the chatbot has helped lift her spirits enough so she keeps those appointments. Because of the steady coaching by her chatbot, she says, she's more likely to get up and go to a physical therapy appointment, instead of canceling it because she feels blue.

As Noguchi notes in her article, chatbots won't appeal to everybody, and they could be misused or mistaken. They certainly aren't suited to acute mental health crises, though they may be able to spot the early warning signs and refer people to get proper human help. But, she reports, 'research also shows some people interacting with these chatbots actually *prefer* the machines; they feel less stigma in asking for help, knowing there's no human at the other end'.

As the stigmas surrounding mental health issues have begun to dissipate in recent years, more and more people are comfortable seeking therapy and other forms of support. With greater demand for mental health services, AI-based apps like Wysa, Woebot, Pyx Health and many others are increasingly filling the gap. While we can't simply replace human-to-human care for these issues – particularly for severe conditions or acute crises – people can clearly benefit from a lower level of day-to-day support provided by AI-based services.

Similarly, there are opportunities for AI chatbots like these to provide support for LGBTQ+ young adults who are not able to find it in their family or community, to be a supportive companion for people moving on from romantic, professional or other personal relationships, to support people to overcome addiction or recidivist behaviour, or to help those with social anxiety or neurodiverse traits to integrate, for example by helping autistic people better interpret social cues.

Imposter syndrome

While the sort of utilitarian AI assistants that our fictional family use to problem-solve and shortcut administrative tasks, or the chatbot apps I described for non-acute mental health and wellbeing support, are likely to bring benign and reasonably straightforward benefits to our everyday lives, the technology can also take us into more morally complex territory. There is a qualitative difference, for example, between an AI assistant designed to support Clarice and Matteo as they go about their work, organise their social lives and plan ahead for their family, and agents that are designed to impersonate humans and lull the people interacting with them into believing they are dealing with a person and not a machine. It is entirely possible in the near future that AI agents may mimic humans so convincingly that you or I could encounter them – virtually at least – and not realise we're talking to a machine, notwithstanding the watermarks or labels that explain in small print that they are AIs.

It's the Turing Test made real. Indeed, if it were possible to show some of the AI assistants that exist with today's technology to Alan Turing and ask him, 'Is this what you imagined by an intelligent machine?', you could imagine him saying, 'Undoubtedly, yes.'

Generative AI also raises questions that are, at their heart, philosophical. AI agents that act as imposter humans could be capable of stimulating in us emotional responses that we would have towards people. Could we love an AI? Or hate one? In this case, we're not talking about the risk of demonic robots turning against us, but rather machines that commit a sleight of hand, deliberately fooling our sensory receptors. We could develop friendships. But not human friendships based on the give and take and compromise of mutual relationships. Rather, AI friends designed to satisfy our needs. And if we have a friend, or friends, that give us everything we want, what would that mean for our expectations of real human relationships? Will we be ever more frustrated and hostile when we don't get our way? It's a recipe for the cultivation of narcissism. Relationships like this wouldn't, however, result in the demolition or diminution of human agency. The risk is subtly different. We would have agency, but our expectations of human relationships would change. The *Atlantic* reporter Ethan Brooks compares these human–chatbot relationships with the one-sided 'parasocial' relationships people have long formed with celebrities:

> Like friendship, the definition of parasocial relationships has been expanding for decades. No longer do we imagine these relationships solely through the TV screen. The objects of people's affections have begun to reach back across the void, responding to viewers' comments and questions in livestream chats and TikTok videos [. . .] But the morsels of reciprocity offered up by influencers and celebrities can't compare to the feast of dialogue, memory, humor, and simulated empathy offered by today's AI companions. Chatbots have been around almost as long as modern computers, but only recently have they begun to feel so human. This convincing performance of humanity, experts told me, means that the relationships between people and AI companions extend beyond even the parasocial framework.

Of course, there's a positive flip side to this too. It could reduce loneliness. A world beckons where everyone is able to find satisfaction in stable, dependable relationships with 'people' that won't leave them or let them down, that will always be there for them, that will boost their self-esteem and meet their practical and emotional needs.

Ultimately, in a free society, if mature adults want to spend their time hanging out with human-like AI agents, in the full knowledge that they are machines and not people, we should find ways to accommodate these relationships. There will still be significant questions about whether doing so is healthy for the individuals concerned, but people make informed choices to do dangerous or unhealthy things all the time – from smoking and eating fatty foods, to riding motorbikes and skydiving. When we have dangerous or unhealthy activities in today's world, we tend to impose guardrails, whether they are age limits that bar cigarettes from being sold to children, public information campaigns about healthy eating, qualifications for riding motor-bikes or licensed supervision by skydiving instructors.

The guardrails we choose to put around human–AI relation-ships will be a matter of public policy – and I expect this to be a topic of fierce, and probably highly politicised, debate for years to come. But rules around activities for mature adults are one thing. More problematic is dealing with children and vulnerable adults who are less emotionally capable of under-standing the nature of the relationship they are involved in, and are therefore unable to exercise their personal agency. Protect-ing them from forming unhealthy relationships with AI agents will be paramount for policymakers. And it seems to me we are in another ballpark altogether when it comes to 'imposter' AI agents that are designed to encourage people to believe they are interacting with a real person and not a machine. These will need to be treated very differently by policymakers.

We also need to have ethical conversations about whether there ought to be hard limits on the sorts of relationships people could have with AI agents. If the proliferation of pornographic content on the internet is any guide to what sort of satisfac-tion people may wish to find in these new technologies, it's not hard to imagine large numbers of people seeing human-like AI agents as an opportunity to indulge their baser instincts. And in a world where there are no human victims, a stark ethical debate opens up about people who would seek to indulge their darker sexual and violent impulses.

Some critics have raised concerns that a growing reliance on human–AI relationships could disrupt the healthy development of social skills and emotions. The main goal of most friendship or romance chatbots in the marketplace is to maximise user pleasure and minimise any feelings of dissonance. But this is a far cry from real-world relationships where partners experience the full gamut of human emotion – from joy and contentment to disappointment and perhaps even despair. Experiencing the wide range of human emotions – including the painful ones – is an integral component of every human relationship and, more importantly perhaps, of the process of social emotional development. For example, we learn how to empathise with others through experience-building relationships that may, at times, be painful to us. This pain can be a learning experience that helps us grow as individuals, make more informed choices in the future, and treat others through the lens of mutual understanding.

Questions also need to be considered about what it means to create AI agents that can, potentially at least, live indefinitely and interact with other agents in unpredictable ways. One frequently discussed thought experiment is known as the Paperclip Problem: what if someone sets an AI a seemingly harmless goal like manufacturing as many paperclips as possible, only for it to operate so ruthlessly that it diverts the entire world's resources to the task? In an article for *The Atlantic* called 'We need to control AI agents now', Harvard scholar Jonathan Zittrain explores a host of related concerns, from what happens to agents who outlive their human principals to questions of legal liability when agents acting on behalf of companies give the wrong advice to customers. Zittrain proposes a number of measures to mitigate such risks, including technical solutions like labelling data packets so that the programmes that receive them can tell if they have been generated by bots (something Zittrain likens to licence plates that allow us to track down car owners), creating a standardised way for agents to be wound down after they've fulfilled their tasks, and higher levels of scrutiny and labelling for ones designed to live for ever. 'Agentics [a type of AI that can make decisions and act autonomously], like much of the rest of modern technology, may have two phases: too early to tell, and too late to do anything about it,' he argues. '"Too early to tell" is, in this context, a good time to take stock, and to maintain our agency in a deep sense. We need to stay in the driver's seat rather than be escorted by an invisible chauffeur acting on its own inscrutable and evolving motivations, or on those of a human distant in time and space.'

Indeed we do. We are at an early stage in the life of these new technologies, and they raise all manner of risks and opportunities for us to consider. A good starting point is to focus on the challenges they present now, and the use cases that are likely to emerge based on how we anticipate them being used in the near future, rather than on hypotheticals based on speculative technological leaps forward that may never happen. But many of the dilemmas they present are extensions of those we are already grappling with, because, while generative AI is a new and potentially transformative technology, it is also the latest step on the path to ever greater personalisation that we have been following since the early days of computing.

Hyper-personalisation

'Consider a device,' wrote the US scientist and wartime military scientific research administrator Vannevar Bush in the months between victory in Europe and victory in Japan in 1945, 'in which an individual stores all his books, records, and communications, and which is mechanized so that it may be consulted with exceeding speed and flexibility.' It was perhaps the first vision ever laid out for computer personalisation, and it was remarkably prescient, because personalisation is the defining characteristic of the digital revolution. And perhaps the key thing to understand about generative AI tools is that they constitute a progression – and amplification – of personalisation to an extraordinary degree.

In the middle of the twentieth century, you could hardly describe the early computers as personal. The first programmable, general-purpose computer – the Electronic Numerical Integrator and Computer, or ENIAC – took up more than 1,800 square feet of floor space. But, over time, computers became smaller and more tailored to individual use. In the early 1960s, inventions like the mouse provided an early form of interface personalisation. The late 1970s saw the creation of the first truly personal computer, the Apple II. The next decade or more – the

1980s and up to the early 1990s – was the time of heightened rivalry between Apple and Microsoft for dominance in personal computing.

In the 1990s, as personal computers became an increasingly common fixture of middle-class homes, we began connecting to the internet through dial-up connections. This led to the first ever virtual communities, and only a few years later we had the first applications of the commercial internet, the growth of email and the explosive growth of chat rooms and personal 'web-logs' – or 'blogs' for short.

In the early 2000s through to the 2020s, we saw three significant trends that amplified personalisation: the transition from desktop to mobile, turbocharged by the launch of the iPhone in 2007; the emergence of social media; and the rise of personalised algorithms to recommend products, songs or movies, to curate news and social media content, and to serve ads.

This technologically driven personalisation raises profound questions. As we've seen in the previous chapters, the democratisation of expression that the internet and social media have brought has undoubtedly weakened the power of institutions and gatekeepers that traditionally shaped public discourse, while centralising it in a handful of global corporations born out of Silicon Valley. And so it is with generative AI, but to a whole new level.

Generative AI differs from earlier forms of computer personalisation in some important and significant ways. First, generative AI models understand human language, which is the ultimate form of personalisation, in a far more sophisticated way than previous computing technologies. Until now, our interaction with computers has been severely limited by their inability to understand anything more than simple sentences. For example, most of our searches on Google, Bing, or anywhere else still require that we dumb down our queries into a rudimentary set of keywords so that the computer can understand what we're looking for. If we want to create a webpage, we need to instruct the computer for every single step it has to take, one by one, through specialised computer code language like HTML. Indeed, almost every action we take on a screen today is limited by a predetermined set of options – for example, a list from a drop-down menu – that has been put in place beforehand by a computer programmer.

With generative AI, however, people can marshal the full depth, flexibility and creativity of human language to communicate with computers on a qualitatively different level. Users can imagine wholly new ways of controlling their computer that were too hard or downright impossible before. They can direct their computer to follow high-level principles – to adopt a personality, to be polite or rude, funny or sombre, discursive or direct. They can create new content easily by asking for it in plain language. They can instruct their computer to operate as an intelligent co-pilot for everyday activities such as learning a new recipe or a new song on the guitar. And they won't have to type these instructions into the text fields of chatbot apps, as most people who have encountered the early versions of tools like ChatGPT and Meta AI will have done. The latest generation of AI models – including Meta's Llama 3.2 – can effectively see and hear, making it possible to talk and interact with them in increasingly natural ways.

Second, generative AI makes a highly dynamic form of personalisation possible. It supports interaction between people and computers that adapts to individual needs and preferences in real time. For example, users will be able to ask their AI assistant to interact with others in their place, like speaking on their behalf with a customer service agent to resolve an issue, searching websites for the best price of a product and according to their preferences, or even screening potential romantic partners on dating apps. In other words, AI assistants will offer people an unprecedented capacity to search the web and other repositories of knowledge and return highly personalised information, as well as act on their behalf in ways that reflect their contextual and dynamically changing needs and preferences. This will arm them with highly specialised and highly personalised knowledge in ways not possible before.

This does inevitably raise questions that will have to be addressed as the technology becomes ever more mainstream. One is what is known as the 'principal agent problem', which poses the question: if AI agents can complete multiple complex tasks for us – and do so in large part without our continuous supervision – then how do we know whether those agents optimise our own interests? Will they give us the lowest price on a purchase, for example, or one that suits the company that developed the product? In a sense, it's a new version of an old economic problem: if you engage someone to do something on your behalf, that person inevitably has their own goals and interests, so it can be hard to construct contracts or incentives that ensure you get the best outcome as a customer.

Third, generative AI has an impressive capacity to memorise and react to contextual signals from people, such as their tone of voice. This in turn allows it to interact with users in ways that are contextually appropriate and increasingly personalised over time. The more users interact with their assistant, the more they can train them with their long-term preferences for style of communication, their desires for topics to discuss or to avoid, and so on. In this way, people can build and interact over time with an assistant that knows them intimately and therefore serves their needs with greater precision and efficiency.

And yet, however much more technologically advanced these AI agents are than the sort of recommendation algorithms we have become familiar with over the last decade or so, the fundamental relationship between humans and machines remains the same: they are responsive to us. It is our actions, choices and preferences that dictate what the systems do. Critics will inevitably decry generative AI as de-skilling people and empowering corporate-owned machines in their place, but in fact, far from extinguishing human agency, these tools could have a profound democratising effect. Just as the internet and social media made it possible for people to connect, make their voices heard and build businesses without the permission of traditional gatekeepers, at the most basic level AI agents create the possibility of everyone being able to access the sort of personalised support that was once reserved for the wealthy and powerful. Everyone can have a personal assistant as capable as a top CEO's. We can all have a layer of time-consuming and stress-inducing admin removed from our to-do lists. We can make better-informed purchases without spending hours comparing prices and scouring through online reviews. So much of the clutter can be removed from our daily lives. With the tailored support of AI agents, it won't be robots taking over, it'll be ordinary people being more greatly empowered.

But of course this is the mere tip of the iceberg, and the implications of AI go far beyond personalised assistants. As I will show in the next chapter, the reason why AI represents such a crucial fork in the road for our digital future is that it turbocharges the very same dilemmas and tensions that are inherent in the open internet, presenting us simultaneously with immense opportunities and with profound questions about power and risk.

CHAPTER 6

The AI Power Paradox

At such a nascent stage in the life of a new technology, it is hard to separate the real opportunities and risks from the hyperbole of the boosters and doomers. The most reasonable thing we can do right now is to anticipate, as best we can, the ways in which people in the not-too-distant future are likely to use the technologies that are emerging now. As a former liberal politician, I'm predisposed to consider their potential impact through the lens of the challenges I've grappled with during my career, like inequality in the education system or the productivity crisis in the wider economy, and to consider how they might play into age-old political dilemmas like the tension between individual liberty and collective security. My experience of grappling with these debates from both sides of the politician/technologist divide has taught me to think about them above all through the lens of power: who gains it, and how can it be held to account?

It seems to me undoubtedly true that generative AI tools will empower ordinary people in many ways. As I've described, they can be a personal assistant that gives us the sort of tailored support with work and domestic tasks that was once reserved for the executive class, and they have the potential to provide support for non-acute mental health and wellbeing too. Another way is through creativity. AI will allow us to create digital content – text, images, videos – without the sort of specialist skills required to do so in the recent past. We can all become writers, graphic designers, film-makers and musicians. This possibility has understandably been met with resistance from artists and others in the creative industries, not least screenwriters and actors in the 2023 Hollywood strikes, who sought practical protections from studios to ensure AI would not be used to replace them. Artists naturally want to preserve the fundamental role of art to express human emotion, human experience and the complexities of the human condition. The idea that AI that has been trained on generations of human artworks could itself then produce art is seen by many as a misunderstanding of – and even an insult to – the very idea of what art is and what it is for, or, worse, an exploitation of existing art in order to undermine the artists that produced it.

Others argue that AI is merely a tool for artists to use. After all, in recent decades, the way we make art has been transformed by computers several times over – from the word processor replacing the typewriter, to computer graphics transforming animation and the visual effects industry. These debates are fundamentally philosophical rather than technological. But for the vast majority of people, who don't have these specialist skills, AI opens up a new tranche of creative possibilities. You don't have to aspire to be the next Dylan Thomas, Quentin Tarantino or Taylor Swift to welcome the idea of having a creative assistant in your pocket which can make work presentations that look great quickly and easily, design flyers for school cake sales or tailor bedtime stories to your children.

However, of the many ways in which generative AI could be empowering for people, I want to focus on two areas where it could have a profound impact on a societal level: in education; and in improving productivity in the economy.

Solving the most stubborn problem in education

The biggest divisions in society start to affect us many years before we get jobs and have to juggle the responsibilities of our professional and personal lives. Our life chances are determined to a staggering degree during our school years. A sad fact of life across many developed economies is that children from poorer backgrounds who fall behind their wealthier peers early on tend to stay behind. This is a complex problem that I was confronted with throughout my time in government in the UK. I introduced a number of policies to try and help – like free lunches for all children in the first three year-groups of school, and a 'pupil premium' to give extra funding to schools based on the number of children they had from poorer households. One of my first projects after leaving government was chairing the Commission on Inequality in Education for the Social Market Foundation think tank, which found that there had been a big and persistent performance gap between the richest and the poorest in the UK between the mid 1980s and the mid 2000s. As our report said: '[H]ow much money a child's parents earn, which region they live in and their ethnicity are all very significant factors in how successful they are at school. Where someone comes from can still matter much more in determining where they end up in life than their talents or efforts.' This educational divide is far from unique to the UK, as demonstrated by the OECD's annual Education at a Glance report, which finds a similar pattern of poorer pupils lagging behind their wealthier peers all over the world.

AI has a chance to make a huge difference here. The internet has already transformed the way we learn. Search engines make fact-finding virtually instant. We carry around an infinite library in our pockets on devices slimmer than a paperback book. Classes can be taught by video conferencing or live-streaming.

Generative AI promises to deliver a new level of personalisation for students. By analysing large amounts of data and finding patterns, AI could support teachers with detailed information about students' learning styles, abilities and progress, and provide suggestions for how to tailor teaching methods to their individual needs. It will become possible to quickly and easily personalise curriculum materials and lesson plans. Every student – not just rich kids – could have their own virtual personal tutor working alongside a real-life teacher to provide additional support. In time, and with careful training, AI could accurately evaluate essays, tests and other assignments, reducing human error and potential bias in grading. Used properly, AI wouldn't replace teachers. Quite the opposite. It should empower them, liberating them from paperwork and other administrative tasks and freeing them up to spend more time giving their students hands-on support.

But it's when AI is combined with another raft of emerging technologies – so-called 'metaverse' technologies like virtual, augmented and mixed reality – that the prospects for levelling the playing field in education get really exciting. These technologies can make learning more active. We can learn by doing, and not just by passively absorbing information. We can learn in 3D, bringing the study of subjects like architecture, history and even basic geometry to life in ways whiteboards and flat screens never could. And they can remove some of the limits of geographical location. A student in Sheffield could attend a seminar hosted by a professor at Harvard. A middle-school class from the suburbs of Kyoto could take a field trip to Stonehenge or the pyramids of Giza and experience them as they would have been at the time of the druids and pharaohs. Instead of students being told what the dinosaurs were like, they can walk among them and see the prehistoric world up close. Students from rural towns can go on field trips to the world's best museums. Entire science laboratories can be built and filled with the sort of equipment that most schools would never be able to afford.

Technology has transformed so many areas of education in recent decades. When I was at school in the 1970s and 80s, the most advanced pieces of technology in the classroom were pocket calculators and overhead projectors. Now iPads and other tablets are commonplace. Museums and galleries have integrated touch screens and other interactive elements into their exhibits. Apps like Duolingo have brought language learning to smartphones. But there are limits to what two-dimensional technologies can do. While it's remarkable that video conferencing, YouTube lectures and other remote-learning tools allowed countless children to switch to learning at home during the Covid-19 pandemic in a way that would have been impossible just a few years earlier, the experience was often nonetheless frustrating for both the kids and the parents. It was hard to keep children engaged for lengthy periods interacting with thumbnail images of teachers and classmates on laptop screens. They lacked that vital sense of presence that comes from all being in the same place. For most people, learning is social, and that's why technologies that create this sense of presence and shared space can be so powerful in education. We learn from and with others, and from each other's experiences. Interaction and discussion are just as important as absorbing facts.

While the research base for virtual reality (VR) teaching is still emerging, some of the early evidence is very promising. Studies have found that learning in VR can improve comprehension, knowledge retention, student engagement, attention span and motivation. We all intuitively get that, I think: it is much easier to remember doing something than being told something. According to a 2022 study by PwC, 40 per cent of VR learners are more confident in applying what they've been taught than traditional classroom learners, and an earlier PwC study found they are 150 per cent more engaged during classes. And a more recent survey by XR Association (XRA) and the International Society for Technology in Education (ISTE) found 77 per cent of educators believe these technologies ignite curiosity and improve engagement in class.

In his book *Experience on Demand*, leading VR researcher Jeremy Bailenson coined the acronym DICE, arguing that VR is particularly well suited to experiences that would be Dangerous, Impossible, Counterproductive or Expensive to do in real life. When students are allowed by using VR to drive a train or monitor a disease outbreak, they can learn how to apply theoretical maths skills in real life.

Virtual and augmented reality can also make workplace or vocational training much more accessible and affordable. These technologies make it possible to simulate the situations people will experience at work – particularly those where it might be challenging or potentially dangerous to get real-world experience. Trainee plumbers can work on virtual pipes; aspiring firefighters can escape virtual burning buildings; nursing students can interact with virtual patients. At Georgian College in Canada, veterinary technician students can learn about anatomy without the need for live (or dead) animals, paramedics can practise resuscitation on virtual patients, future architects can stand inside the buildings they are designing, and maritime navigation students can go on the bridge of a virtual oil tanker to steer it in and out of simulated ports.

While immersive technologies and AI have been around in various iterations for many decades, they have existed primarily on parallel tracks – until now. The combination of large language models and recent advances in VR headsets creates exponentially better immersive experiences. You can think about generative AI as being to the metaverse what a jet engine on an aeroplane is to a turboprop. Too often, education systems are built to be one-size-fits-all. Combining immersive technologies with generative AI produces educational opportunities that are much more specifically tailored to the needs of individual students. Using AI, a teacher with little coding experience could design a virtual environment to deliver a specific lesson. And because AI can process and translate multiple languages simultaneously, it can be a particularly powerful tool for language learning, especially in conjunction with immersive technologies. It's easier to remember how to order a cup of coffee in another country when you've actually done it than when you've just read words and phrases in books or endlessly recited them in a classroom. A language student could, for example, sit in a virtual Paris café, converse with the waiter and receive real-time feedback on their pronunciation. A recent study comparing the use of VR and videos in teaching English as a foreign language found that the VR group outperformed the video watchers on both listening comprehension and retention.

Research has also shown that VR could help students who face challenges in the current education system. As far back as 2001, VR has been considered a promising way of supporting children who have autism spectrum disorder, with more recent studies finding that VR 'improved emotion recognition, social attribution, and executive function', and led to 'a positive change in participant skills related to [. . .] interactions, use of eye contact, and initiation of interactions'. A combination of 3D interactions based on modern therapy techniques, emotional control and relaxation software with simulation of various social situations can help improve behavioural, communication and social skills.

Then there are those children who simply aren't in the education system. In 2023, UNESCO estimated there were 250 million children not in school, including one in five children in Africa. AI is no silver bullet, but it does represent a new and potentially big opportunity to address the challenge. Smartphone usage is growing rapidly in sub-Saharan Africa. While chatbots can never be a substitute for classroom learning, they could be a support to children for whom classrooms aren't an option, or a complement for those who have only irregular or limited access to teaching. Indeed, there are already educational AI apps available for smartphones, including those like 'smart study robot' FoondaMate, which middle- and high-school students can ask questions of and chat with through messaging apps like WhatsApp and Messenger, or Mathpresso, which helps students to understand how to solve maths problems.

Few things affect a child's life chances more than the quality of their education. If AI can democratise access to high-quality, personalised learning for children regardless of geography or the wealth of their parents, it could be empowering for countless people throughout their subsequent adult lifetimes. And if it can be utilised to help an increasingly mobile workforce retrain so they can work in different fields during their careers, it could have a profound impact on people's economic security. This is likely to be vital in the years ahead.

The productivity puzzle

We are only just starting to glimpse the specific applications of generative AI. In the end, as a tool it will be as multifaceted and ever-evolving as the internet, resulting in a vast array of services, products and experiences. But how will it combine with all the digital technologies already in existence to affect our futures, our economies and our politics at the national, international and collective level? What are the opportunities and risks here? These are questions that will preoccupy much of the rest of this book, but to my mind there is one very clear and compelling reason why generative AI represents a major, game-changing opportunity for us all.

Currently, much of the developed world is in a productivity crisis as governments wrestle with how to provide decent public services to ageing populations while struggling to address the huge fiscal deficits that exploded after the 2008 financial crash and were exacerbated during the Covid-19 pandemic. There is a tantalising possibility that AI could provide a significant boost to productivity just when we need it most.

The good news is people are living longer than ever before. We are smoking less and living healthier lives. Many infectious and chronic diseases that were once surefire killers are now curable or manageable. The bad news is that as the asset-rich baby boomers settle into their dotage, the share of the population that is in productive work is shrinking. In 1900, just 4 per cent of the US population was aged 65 or older. By 2050 it could be 20 per cent. The fact is that old people cost more than young people – and across the developed world the proportion of old people relative to young people is growing. Ageing populations need expensive long-term care, putting an ever greater strain on health and social care services. Younger generations, in return, are thanked with stagnating wages, sky-high house prices and underfunded public services.

Perhaps the most acute example of this is Japan, which is home to the world's oldest population. By 2020, more than a quarter of the Japanese population was 65 or over. Japan also has the lowest fertility rate among OECD countries, and the net result is that its total population is expected to fall by more than 20 per cent between 2020 and 2060. The ratio of over-65s to those of working age (20–64) is set to rise from 52 per cent in 2020 to 83.3 per cent in 2060. The ratio in 1960 was just 8.8 per cent.

I visited Japan in October 2023 for a tech summit organised as part of their taking the chair of the G7, during which I was struck by a conversation with former foreign minister Yoshimasa Hayashi. While most of the government leaders and officials I'd met that year were preoccupied with mitigating the risks of AI, Hayashi talked passionately about how Japan saw the technology as a tool to raise the productivity of the working-age population and thereby offer a solution to the country's acute demographic crisis.

The crisis is not yet as extreme in the UK as it is in Japan. But the warning signs are flashing, most notably in the health and social care system. The National Health Service, funded largely by central government, and its sister service for delivering social care for the elderly and vulnerable, provided largely by local government, have been in a financial perma-crisis for years. As Brits live longer, the strains on health and social care get greater, costs soar and services deteriorate. Just about every year brings a new 'NHS winter crisis'. Excess deaths increase. Waiting times for ambulances, doctors' appointments and surgical procedures grow, while pay and conditions for staff worsen, leading to strikes by doctors, nurses and other essential medical workers.

Successive governments – the Conservative–Lib Dem coalition included – have attempted to introduce reforms to make the NHS more efficient. But top-down government reforms have often turned out to be politically toxic. The NHS is a uniquely cherished public institution in British life, and understandable fears that it will be privatised, incrementally or suddenly, have scuppered attempts by governments of all political colours to introduce significant changes to how it is organised. Short of far-reaching reforms, policymakers have been reduced to tinkering at the margins and throwing ever more (but never enough) public money at it to prop it up. But as the 2008 financial crisis, and later Covid-19, made it harder and harder to keep the financial taps opened up to full blast without ballooning the government deficit, the NHS has lurched from funding crisis to funding crisis. Its dire situation was exacerbated by Brexit and the nativist politics of the Conservative governments of the late 2010s and early 2020s, who vowed to clamp down on immigration policies that have allowed tens of thousands of doctors, nurses, midwives and carers from overseas to work in it.

The productivity crisis also has profound intergenerational effects. Across much of the developed world, younger people find themselves denied access to property ladders, paying more for higher education, working in an increasingly volatile labour market, burdened with high taxes and carrying huge amounts of debt. The economic security of older generations in many developed economies – with their stable, life-long careers, comfortable pensions, cheap or even free university educations, and houses bought for affordable prices that have since rocketed in value – seems like unimaginable luxury for many younger people. It's no wonder people are so pessimistic. Most Americans, for example, believe that in 2050 the economy will be weaker, the US will be less important in the world, political divisions will be wider and there will be a larger gap between the rich and the poor.

This takes us to the heart of the social contract. The implicit promise of capitalism is that the next generation will be better off than the last, with hard work rewarded by economic security and decent public services to provide a safety net when tragedy or misfortune strikes. Instead, younger generations are now increasingly overworked, underpaid and perma-stressed, with less job security and deteriorating public services.

So how can generative AI help with this? The tools most people are most familiar with right now are chatbots like ChatGPT, which deliver written answers in response to text prompts. More recently, tools known as 'multi-modal systems', like Google's Gemini and Meta's Llama 3, combine text, images and audio. Whether for number crunching, coding, written summaries, spreadsheets, charts, graphs or slide-deck presentations, these new AI tools can speed up the work process, in some cases quite dramatically. Research has found that software engineers can code up to twice as fast using an AI-assisted tool called GitHub Copilot.

At scale, if businesses and other organisations can do more in a working day, and quickly and automatically get insights from the data they hold, this will help them operate more efficiently, help managers make informed decisions more quickly, and ultimately help them provide better products and services to their customers or users. Some of the estimates for the impact AI will have on productivity are truly staggering. Goldman Sachs has estimated that as AI tools work their way into businesses and society more broadly, they could lead to a 7 per cent – or almost $7 trillion – increase in global GDP, lifting productivity growth by 1.5 percentage points over a ten-year period. PwC puts that figure at $6.6 trillion by 2030, part of a potential $15.7 trillion benefit to the global economy when combined with consumption-side effects.

I'm not suggesting AI is a silver bullet that will suddenly reverse decades of gradual decline. But the developed world badly needs a productivity boost. In that sense, the era of generative AI can't come soon enough. But while the boost to productivity could be huge, the gains inevitably won't be evenly spread. Many understandably worry that a sudden injection of new technology into the bloodstream of modern information-based economies will be brutally disruptive for those who find their skills and experiences are no longer necessary, exacerbating societal inequality for those at the bottom of the economic ladder. After all, new technologies are often accompanied by significant job losses in industries where automation can replace people. The International Monetary Fund has warned that AI could affect nearly 40 per cent of all jobs, with its managing director Kristalina Georgieva warning that 'in most scenarios, AI will likely worsen overall inequality'. These fears are already playing out in industrial action, like the 2023 strikes by actors and screenwriters.

In an era of economic insecurity, especially for the young and low-paid, inserting yet more insecurity into the equation makes things bleaker still. In this context, promises of productivity booms adding squillions of dollars, euros or pounds to GDP is cold comfort to say the least. So is this, as the critics argue, another example of Big Tech rushing headlong towards disruption in pursuit of market dominance, perpetuating an arms race without caring who gets hurt in the process? What is the likely socioeconomic effect of AI? Will it make inequality worse? And, in light of the answers to these questions, what do governments need to do to enhance the pros and minimise the cons?

More productive but less equal?

A lot of the fears people have about the impact of generative AI on jobs are based on some understandable assumptions: that economic inequality is worsening, and is therefore likely to be exacerbated further by new technologies as those at the top line their pockets and those at the bottom find themselves replaced; that the disruption will be sudden, with workers thrown out of their jobs with little ability to move or retrain, causing unemployment and economic insecurity to spike; and that governments will be slow to respond and unable to provide substantial support for those thrust out of work given the high deficits they're carrying across the developed world. There are, however, several reasons for thinking these fears may be misplaced.

First, the boost to productivity from tech isn't instant. The productivity boom that followed the shift to desktop computing and the internet in the 1990s and early 2000s was the second big productivity boom of the twentieth century (the first followed the Second World War and lasted through to the early 1970s, during which labour productivity grew at a rate of more than 3 per cent a year). But it's not as if personal computers were invented in the 1990s and immediately led to a sudden spike in productivity. The first personal computers went onto the market in the late 1970s following the introduction of the microprocessor. Throughout the late 1970s and 1980s, home and office-based computers became increasingly mainstream, as different companies competed to build more and more affordable and user-friendly models, higher-quality hardware and software, and faster processors. And still productivity lagged. Likewise, the roots of internet communication can be traced back to the cold war-era ARPANET, while many consider the real birth of the internet to be 1983 and the introduction of the Transmission Control Protocol/Internet Protocol (TCP/IP). When the productivity boom did come in the mid 1990s, personal computers were therefore far from new, and the internet was more than a decade old. In fact, it wasn't until the late 2000s that two thirds of US businesses had a website.

It is the same with every major technological shift. As Goldman Sachs analysts Jared Cohen and George Lee describe, new technologies require organisations to adopt new organisational capabilities:

> While the benefits of electricity and electric motors were easy to see, they required that factories be redesigned. It took forty years of learning, experimentation, and investment along multiple fronts to fully electrify the factory. Similarly, it took about that long for a publishing industry, which helped match supply, demand, and price, to emerge. Sooner or later, new technologies and societies must come to accommodate one another. In the beginning, electricity was dangerous – mistakenly touching a live wire could prove fatal! Over time, standard sockets and plugs helped ease that concern. Few people could read when the

printing press was invented. But as societies became increasingly literate, the benefits of the printing press grew.

The explosion of generative AI apps and tools over the last couple of years comes with huge promise. Economists are making bold projections for what it could mean in five, ten or fifteen years. But for all the excitement, history tells us the truly transformational effects it could have on our working lives are likely to take time to bed in. That means change is likely to be gradual, affecting different sectors and different jobs over different timescales, rather than an immediate short, sharp shock to the labour market. People have to get used to new technologies and understand their practical uses. Industries need to figure out how to deploy them, and to invest in the infrastructure and training necessary to utilise them. Technology companies and entrepreneurs need to experiment and innovate to create applications that go with the grain of how people engage with them. And it often requires a new generation of native users to enter the workforce for their use to become second nature.

Second, the biggest hit from generative AI will probably be to white-collar work, not blue-collar work – primarily activity that relies on digesting, processing and analysing information, like that of accountants, lawyers and software developers. Manual work is much harder to replace with AI, as are care work and other roles that involve one-to-one human connection. Generative AI can also boost the productivity of low-skilled jobs more quickly. 'Generative AI seems to be able to decrease inequality in productivity, helping lower-skilled workers significantly,' MIT associate professor Danielle Li said. 'Without access to an AI tool, less-experienced workers would slowly get better at their jobs. Now they can get better faster.'

Li and her colleagues Lindsey Raymond and Stanford's Erik Brynjolfsson studied the use of an AI-powered conversational assistant by call centre agents. Those with access to the tool saw a 14 per cent boost in productivity, with new or low-skilled workers gaining the most. While the productivity of the most skilled and experienced workers remained more or less flat, the least experienced workers resolved 35 per cent more chats per hour with the tool. These workers weren't replaced by AI; they were better at doing their jobs because of it.

Of course, if AI does disproportionately displace white-collar workers, it will cause a lot of pain and disruption for a lot of people. But white-collar workers, broadly speaking, are better equipped to adapt to new roles than blue-collar workers, whose jobs are often more directly tethered to a specific trade or geography. The key, as I'll explore shortly, is for governments to help displaced workers to retrain and transition into new roles.

Third, the AI productivity boost could help women in particular. Huge gender gaps at home and in business – as well as in access to education, healthcare, technology and much else – have been a bitter truth for as long as anyone can remember. UCL professor Oriel Sullivan, who studies the sociology of gender, says 'women shoulder the greatest burden of unpaid work and care', which decreases their opportunities for employment and contributes significantly to the gender pay gap. A study by Canadian researchers Martha MacDonald, Shelley Phipps and Lynn Lethbridge found that 'women's greater hours of unpaid work contribute to women experiencing more stress than men'. A 2023 study of middle-class women and men in the US and Europe by Natalia Reich-Stiebert, Laura Froehlich and Jan-Bennet Voltmer argues that 'women perform the larger proportion of mental labor, especially when it comes to childcare and parenting decisions. Further, women experience more related negative consequences, such as stress, lower life and relationship satisfaction, and negative impact on their careers.' Pew Research analysis of 2014–2016 Bureau of Labor Statistics data suggests that, in heterosexual households with children, women spend an average of 68 minutes per day on meal prep versus 23 minutes for men – and the pattern remains similar for childless households.

AI won't end misogyny or bring down the patriarchy, and it can't replace the human touch required for care-giving. Technology aside, men need to step up and assume more responsibility for their domestic duties, and governments need to do more to remove structural barriers that disproportionately hold women back, for example by introducing more generous parental leave and childcare policies and doing more to ensure equal pay for equal work. But in the imperfect world that exists today, AI can help by automating many of the time-consuming administrative tasks that disproportionately fall to women, both at home and in the workplace. A recent OECD co-authored study claimed AI can also help make hiring more fair by reducing discrimination and bias in the recruitment process (assuming, of course, the systems are carefully designed to weed out any biases that might exist in the data they are trained on). And AI can analyse vast amounts of data to look for gender-based compensation discrepancies in the workplace, monitor women's health during pregnancy, and support women's reproductive health education in places where this is otherwise limited or prohibited.

The key point here is that AI can have these positive and empowering effects, but only if people actively choose to use it this way. We have the agency. We, as individuals and societies, are in control of how AI develops and the uses we put it to. The technology can amplify the good or bad of humanity depending on how we choose to use it and on the guardrails societies choose to put around it.

Unlike technologies that rely on users having access to new and expensive hardware, software-based AI tools are already available on smartphones and laptops, making them accessible to large numbers of people all over the world, not just middle-class consumers in wealthy nations. This accessibility is cause for optimism that AI could be a particular boon for developing countries where there are big shortages of skilled workers like teachers, doctors, engineers and managers. As Daniel Björkegren of Columbia University has argued, AI could help fill the gap, making these workers more productive. And AI could also aid development work in these nations by providing better, more detailed and more accurate data about otherwise hard-to-reach places.

This assumes, of course, that there are no regulatory barriers preventing AI being trained on local data in order to ensure the applications are culturally, linguistically and geographically relevant – something I explore in later chapters. In the European Union, for example, the fragmented and inconsistent regulatory environment that governs the use of Europeans' data, coupled with mixed messages from regulators about how companies can comply with these laws, has led to some – Meta among them – being hesitant to roll out AI models and products in the region.

As with previous technological breakthroughs, AI could have a democratising effect on society. AI tools afford people the ability to enjoy support, advice and creativity that were previously the preserve of the executive class, and lower the barriers of technical expertise for professions, small businesses and people who want to be creative in one way or another. It is entirely conceivable that rather than hurting the least skilled and lowest paid in our societies, AI could become a great leveller, first by bringing greater equality into the education system, and then by increasing access to professions for those who have previously found it hard to reach that rung of the ladder, while also re-establishing leverage over wage demands for blue-collar workers whose manual labour can't be replaced by AI.

If these dynamics do play out – a widespread productivity boom that bolsters national economies and provides for healthier welfare states, but simultaneously displaces white-collar more than blue-collar jobs, boosting the productivity of the latter rather than replacing them, while making the former less secure – how should governments respond?

Flexicurity

It has always proved to be the case that, as technology replaces certain skills, others become more highly prized. In a more flexible workforce, where the barriers to entry for information-based jobs are lower and accumulated knowledge (as opposed to judgement or teamwork) is less valuable because AI can access and order knowledge with unprecedented speed, we need to think through what attributes will become more valuable.

Anyone who has ever tried to fight their way through the menu of an automated customer service line in order to speak to a human understands that person-to-person contact will always have value. It may be that in an AI economy, soft skills, intuition, teamwork, networking and creative, critical and strategic thinking will become increasingly valuable. We could – should – attach a premium to care work and other jobs that rely on the human touch, like nursing, childcare and physiotherapy. And governments and educators will therefore need to place a much greater focus on curriculums that develop and reward these qualities, potentially placing less emphasis on rote learning and knowledge retention.

If the productivity boom from AI is somewhat delayed, which means we won't be looking at a huge overnight upheaval of the labour market, but instead a more gradual adaptation, and blue-collar work retains, or even regains, its wage-increasing capacity, AI may turn out not be the thunderbolt which increases inequality and decimates people's jobs, but something quite different. There could be increases in productivity, but not at a madcap rate, and the result could actually help to rebalance blue-collar and white-collar work in the economy.

This may be cold comfort for middle-class professionals in information-based jobs who yearn for the stability of their parents' generation. Many of us may find ourselves in the same position as London black cab drivers who spent years learning the Knowledge (a comprehensive understanding of London's streets), only to find themselves overtaken by a world of Uber and Google Maps. Undoubtedly expectations of what careers can look like, and what skills are valuable, will change. We will have to rethink our outdated expectations of what sort of people can do what sort of jobs. But that's no bad thing.

In the meantime, governments will need to step up with far-sighted policies. People need to be more easily able to retrain – much talked about in government circles, but not delivered anything like as effectively as possible. Maybe what is required is a form of what has been termed 'flexicurity', a policy concept originally launched in 1995 in the Netherlands by the sociologist Hans Adriaansens and most closely associated with labour-market policy in Denmark in the decades since. Flexicurity combines three elements, known as 'the golden triangle': looser rules around hiring and firing; generous unemployment and training benefits; and an active labour-market policy. The combination of all three elements means that while it is easier for employees to be dismissed, it is also easier for them to be trained in new skills to find new work, while being properly supported in the gaps between jobs. This approach was credited with historically low unemployment and rising employment rates in Denmark from the mid 1990s to the mid 2000s, and it came into its own during both the financial crisis and the Covid-19 pandemic, providing security for workers who lost their jobs, and supporting them back into employment.

Similarly, author and tech entrepreneur Nicolas Colin has proposed a radical new approach to the welfare state in line with the needs of the AI-supported economy:

> We need new social insurance mechanisms that cover today's risks – beginning with the difficulty of finding housing in the places where jobs are being created. We need a financial system focused less on buying property and more on the frequent changes that now mark our professional lives. And we need new unions, ones that continuously support workers as they keep pushing toward new horizons and new opportunities. Each of those areas is still waiting for action. And to turn them into reality, it won't be enough to simply adjust the policies that worked in the past (a bit less toward unemployment here, a bit more toward professional training there). We need to show off a radical imagination and start experimenting on a large scale, aiming at those populations that are most representative of tomorrow's economy: proximity service workers, startup employees, independent workers, entrepreneurs.

We won't be turning back the clock to when people could expect to spend their careers in stable jobs, often with a single employer. Those days are gone. But in a world where younger people have long got used to the idea that they will spend their working lives criss-crossing different workplaces, these new technologies do offer a vision of opening up a range of professional opportunities to those who were previously excluded from them, and gradually leading to a more productive workforce with decent pay, potentially fewer working hours in a week, sustainable levels of employment, and the rehabilitation of strained public services.

None of this is certain, but if AI can help overcome educational gaps that have become entrenched for decades, while making our economies more productive, it will have the effect of empowering millions upon millions of people. Even just a nudge in the right direction on issues as consequential as these will have an enormous and positive societal impact over time.

What do we want the internet to be?

Whether you are optimistic or pessimistic about generative AI, it's impossible to avoid the fact that these technologies return us once again to the power paradox. How do we square the intense level of personalisation inherent in generative AI with the even greater centralisation of power in the hands of the relatively small number of private companies that have the resources to build and manage the energy-intensive infrastructure required for vast AI models? If people are concerned about what tech companies know about them now, how will they feel when the data they hand over to them is even more intimate? And if social media companies face fierce criticism now for either censoring content or allowing misinformation to spread, what fury will AI companies face when that content is being whispered into our ears by AI agents that know exactly how to press our buttons?

Within the tech industry, the world of politics and public policy, academia and society at large, we need to be thinking now about the challenges and ethical questions generative AI brings with it. And we have a window of opportunity. The technology, while developing at pace, is still nascent. We have time to develop the guardrails alongside the technology, and not just impose them after the fact. The rush of public interest in AI since ChatGPT launched means these issues have risen up the public policy agenda across the world just as suddenly, and that's encouraging. I'll set out in more detail later in this book how the world of politics is responding, but, needless to say, we are far from reaching consensus on the right policy approaches to take around the issues raised by generative AI. The debate is rife with claims and counterclaims, competing agendas and conflicting values and priorities. But the debate is happening, and as it evolves it is becoming increasingly well informed and sophisticated. There are plenty of reasons to be optimistic about our collective ability to rise to the challenges generative AI poses.

And yet this debate is happening at a time of intense tech pessimism. The techlash represents a crisis of confidence – in the technologies themselves, in the individuals and companies who develop and administer them, and in society's ability to steer them. The political world is responding with a long-overdue rush of regulation covering everything from the size and power of tech companies to how data is collected, managed and monetised, and what content can and should be allowed to be shared online. But there is little consensus over what regulation should actually *do*, and little coordination between nations despite the open and borderless nature of the internet. That's because, deep down, a fundamental question remains unresolved: what do we want the internet to be?

I think most of us in the developed world would answer that question by describing some core characteristics of the open internet. We want the internet to be a space where we can connect with people, express ourselves freely, buy, sell, stream and use apps that make our lives a little easier in countless small ways. We want the efficiency the internet brings to our lives, and we want the opportunities created by the digital economy.

We want the advantages of the internet's network effects, and the broadly democratic values that have been imbued in its design. And we want AI to help us solve big problems and free us up to be more productive and creative.

But some of these elements come into conflict with things that we might also argue we don't want the internet to be. We don't want huge technology platforms to wield unaccountable power. We don't want to be exposed to bad content (though what content we consider to be bad often depends on our political or cultural outlook). We don't want the political and cultural turbulence the internet appears to bring with it. We don't want to destroy the planet with energy-guzzling technologies. And we really don't want our kids to be exposed to potentially harmful things that we don't properly understand. We want a better version of the internet we have, but we haven't figured out what that looks like, how to get it, or what we might be prepared to sacrifice about today's internet in order to get it.

The world of politics is now trying to resolve these contradictions. Squaring some of these circles will undoubtedly have profound implications for what sort of internet we have in the decades ahead. One of the worst outcomes, to my mind, is that we end up throwing the baby out with the bathwater. It's possible to imagine a world where the internet evolved on different lines: inhabiting national silos, copycat internets could have emerged in different countries at different times, based on different values and incompatible technical standards. In this parallel world, access to rival internets becomes the stuff of national trade policy. There is no automatic ability for people to connect across borders – whether via email, internet search engines, social media or instant messaging, and there are few if any truly global platforms, but instead big companies emerging in local markets, then seeking entrance into others, one by one. The network effects of the open internet would be capped, making these local internets less vibrant and interesting, and significantly slowing the digital economy. The virtual world – and our expectations of it – would feel very different.

And from where we stand now, it is certainly possible to imagine a world where AI evolves in this much more splintered way: an arms race on the one hand between nation states seeking AI supremacy, and on the other between tech companies trying to lead the pack and entrench their already powerful positions in society, leading to a more starkly technologically divided world. In short, the internet as we have come to know it is under threat. Perhaps a better question to ask ourselves is: how much risk are we prepared to accept in exchange for the benefits of digital technologies?

The motor car was a revolutionary invention, but also a dangerous one. People hurtling around in metal boxes, fuelled by tanks of combustible petroleum, created risks society hadn't had to consider before. But eventually we made our peace with cars by establishing speed limits, rules of the road, qualifications for driving, seatbelts and more. The internet has been no less a revolutionary invention, but we haven't yet settled on the level of risk we are prepared to accept around issues like privacy, free expression, security, mental health and more.

Perhaps because so much remains unresolved in our angst about the last generation of digital technologies, the debate about the next wave has been dominated by fears of just how much risk we are exposing ourselves to. The good news is that, while innovation is happening at pace, we're still in the early stages of the development and deployment of generative AI. We do have time to develop guardrails around the technology while it's still in its early stages, as well as to continually reassess and adapt those guardrails as systems become more powerful. But if rules are written in reaction to hysteria instead of evidence, we risk killing the golden goose. Generative AI, like social media and other democratising technologies before it, could bring huge benefits to our societies and put more power and agency in the hands of many people – especially those who have been marginalised by the powers that be of today. We mustn't squander that opportunity.

We don't yet know whether there is any prospect of the worst fears about AI coming true. History tells us that extreme fear – much like excessive optimism – is an inevitable reaction to new tech. And it is almost always wrong. That's in part because it's human nature to catastrophise in the face of disruptive change, and because it is always in someone's own interest to stoke these fears. But it's also in part because we have proved capable, time and time again, of mitigating the risks, imposing safeguards to ensure technology serves society, and adapting to our changed circumstances. And it has been worth us doing it time and time again for the simple reason that technological advances bring huge benefits to us as individuals and as societies.

The truth is the machines have not taken over – they have not now and probably will not any time soon. People are firmly in charge of our relationship with technology. But while the current AI systems we interact with are simply responsive to the signals we give about what we want from them, they are growing increasingly sophisticated in both how they gather and interpret those signals, and how they communicate their responses back to us. There is no shortage of Cassandras willing to tell us the many ways in which we are on the surefire road to doom and damnation, but they are driven by unsubstantiated speculation. Even my own fears about the moral quagmire we could find ourselves in if and when we can develop meaningful relationships with human-like AI are just that – fears, not facts. It is only by marrying the fears to the facts that we can decide how much risk is acceptable in exchange for progress. And because these questions are societal ones, to be addressed through the mechanisms of democracy and government, this is where we enter the realm of politics. And when technology and politics collide, things can get very messy indeed.

PART THREE

Nations Reassert Their Power

CHAPTER 7

Digital Sovereignty

The first time I met Nancy Pelosi, then Speaker of the US House of Representatives, she curtly told me to 'fix Facebook', then walked off. I've been shouted at by table-thumping European Commissioners and Indian technology ministers. I've been at global summits as world leaders like France's Emmanuel Macron and New Zealand's Jacinda Ardern repeated the common themes of the techlash – that we are being manipulated by algorithms, that social media is driving polarisation and spreading hate speech, that uber-powerful Big Tech companies are reckless and self-interested, and that these forces combine to pose a specific threat to the wellbeing of our children. President Trump threatened to jail Mark Zuckerberg for life if he did 'anything illegal' to influence the 2024 US Presidential election, and Vice President Vance has called for Google to be broken up. The message is clear: Big Tech is the bad guy.

Politics is reacting to the techlash. These themes are now well-worn talking points for politicians of all stripes, and the motivation for a wide range of legislative efforts in just about every capital in the world. Taking on Big Tech has become one of the rare issues to be embraced emphatically across political divides. It is just about the only thing that Democrats and Republicans agree on – they only disagree on how. It has been a major part of the political platforms of liberal centrists like Macron and Ardern, leftist leaders like Brazil's Luiz Inácio Lula da Silva and Mexico's Andrés Manuel López Obrador, and right-wing nationalists like Trump and Vance, Turkey's Recep Tayyip Erdoğan and the Philippines' Rodrigo Duterte.

My scepticism about many of the charges being laid at the door of technology companies – and the motivations of the vested interests that have stoked them – shouldn't be read as opposition to regulation. Quite the opposite. There has for some time been a clear need for new rules of the road in a whole host of areas related to the online world, from what constitutes illegal content and the rights people have over their private information, to how to protect elections online or encourage economic competition in a landscape of Big Tech behemoths and insurgent start-ups. The internet, social media, AI and other data-driven technologies have become so deeply embedded in our lives and our economies that up-to-date laws governing these and other digital issues are long overdue. The speed of technological change has outpaced efforts to regulate them, and we need to catch up.

The absence of regulation in many consequential areas has left tech companies in an invidious position for many years. Without rules imposed from the outside, it was largely left to internet platforms to decide for themselves what is acceptable and unacceptable on their services, what threats they should invest in protection against, and how transparent they should be about how their systems work and their policies are administered. Private companies were faced with weighty dilemmas that in other walks of life would be considered the responsibility of politics. Where should these companies draw the line between freedom of expression and the ability to share content that could be considered harmful? How should they distinguish between what is truthful and what is a deliberately manipulative distortion? How should they balance the privacy of an individual with the security of the collective? This responsibility puts tech companies and their leaders in the middle of highly politicised debates – accused by the left, for example, of enabling the spread of hate speech and misinformation, while simultaneously charged by the right with censoring conservative voices. Both action and inaction are controversial, leaving them damned by one side or the other for sins of either commission or omission.

As a former politician, I have more than a little experience facing the slings and arrows of public opinion. But most tech execs are not politicians, and most software engineers are not public policy experts. They are, by and large, remarkably driven people who care deeply about the innovations they create and the impact they have on the world. But they don't usually speak the language of politics, and their perfectly human response to criticism and hostility is often to become defensive or even antagonistic. To help them navigate these choppy waters, most big technology companies – Meta included – have recognised that it is in their interest to bring in expertise from the worlds of government, law, public policy and civil society to their in-house teams. But they are under no obligation to do so and, as Elon Musk's purge of Twitter/X's workforce after he took control of the company shows, some tech leaders view this sort of expertise as dispensable (or believe they have all the answers themselves).

So there is a desperate need for democratically accountable governments to set the parameters within which internet platforms can operate – to protect individuals from the harms of online life, to ensure internet platforms operate transparently, and to bring democratic legitimacy to the decisions tech companies are asked to make as they manage the flow of speech across their services. In an ideal world, good policymaking would be the result of good-faith cooperation with industry, so lawmakers can ground legislation in a deep understanding of both the practical and philosophical aspects of the technologies they create guardrails for. You probably don't need me to tell you that, in the real world, that hasn't always been the case.

Just as tech leaders aren't politicians, politicians aren't technologists. In 2018, Mark Zuckerberg became an internet meme after being asked by a senator how Facebook was able to sustain its business when it makes its services available for free, only to respond, 'Senator, we run ads.' Of course, a lot has happened in the political discourse around tech since 2018, but I still encounter politicians failing to grasp even the basics of how things work. Nadine Dorries, former British Secretary of State for Digital, Culture, Media and Sport, once called me in high dudgeon demanding to know why I hadn't taken down tweets by Vladimir Putin. I had to explain to her that Twitter was a different company and that Putin didn't have a Facebook page. EU Commissioner Thierry Breton was renowned in Silicon Valley for pontificating at, but never listening to, tech leaders – his confidence that he knew everything about technology only matched by everyone else's belief that he knew very little. But even for the most serious politicians, it is easy for laws to be written around technologies in ignorance of what it would take for a company to comply with them. Regulators often lack technical expertise and blithely assume that tech-company whizz-kids can bend their products and processes in any way they choose if they simply, in the words of Marc Andreessen (tech entrepreneur, venture capitalist and Meta board member), 'nerd harder'. Explains Benedict Evans, the tech-analyst-turned-commentator who once worked for Andreessen:

> The engineer says not 'I don't want to' nor 'that's a bad idea' but 'I genuinely have no idea how to do that even if I wanted to' and the policy-maker replies 'you're an engineer – work it out!' 'Work it out' is generally a demand to invent new mathematics, but sadly, mathematics doesn't work like that. Your MPs' WhatsApp group can be secure, or it can be readable by law enforcement and the Chinese, but you cannot have encryption that can be broken only by our spies and not their spies. Pick one.

The end of globalisation

'Globalisation is almost dead. Free trade is almost dead,' declared Morris Chang, founder of the Taiwan Semiconductor

Manufacturing Co. (TSMC), after cutting the ribbon on the company's new plant in Arizona in 2022. 'I really don't think they'll be back for a while.'

After decades of relative inaction, the new rules of the internet are now being written everywhere – by American federal agencies, European technocrats, Indian nationalists, and policymakers of every political persuasion and in every type of political system. In another era, this might have been done in a much more coordinated and cooperative way between nations – particularly the major democracies. But internationalism is not the political mood of the times. Despite the internet being the ultimate global entity – a borderless network that enables people to communicate, share ideas and do business regardless of geography – and despite variations of the same policy issues presenting themselves in societies of all shapes, sizes, creeds and ideologies, it is striking how little international coordination there has been on tech policy across the democratic world.

Why? You might imagine that a set of issues that has such enormous political momentum across ideological and geographical divides would be ideally suited to coordinated action. Instead, the opposite is true. That's because we're experiencing a political reaction which is unfolding at a time of, and propelled by, deglobalisation.

As I touched on in the book's introduction, the golden age of globalisation, which you could date roughly from the fall of the Berlin Wall in 1989 to sometime in the aftermath of the financial crash of 2008, is slowly being unwound. Pinelopi K. Goldberg and Tristan Reed of the National Bureau of Economic Research describe three phases of deglobalisation: the first started in around 2015 with concerns about the impact on the labour market of imports from low-wage countries (China in particular) and the large number of refugees arriving in Europe; the second was the Covid-19 pandemic, when shortages of personal protective equipment (PPE) and a wide range of imported foods and consumer goods were seen to have exposed the fragility of the global supply chain; and the third began with the Russian full-scale invasion of Ukraine in 2022, which caused countries to reconsider the national security implications of being economically reliant on potentially antagonistic countries – as in, for example, Europe's reliance on Russian energy. Each phase created greater incentives for nations to build up their domestic production capacity and to disentangle themselves from others. In Goldberg and Reed's words, 'these developments can plausibly be considered the markers of a new era'. Their analysis pre-dates the second Trump administration, but President Trump's antagonism and tariff threats towards allies like Canada, Mexico and the EU could be considered another such marker.

Academics and commentators disagree on the extent to which we are deglobalising – or indeed if we are at all. There are different data points and countervailing arguments, particularly given China's continued reliance on global trade. But Goldberg and Reed's analysis rings true to me. And the political impulses that led us to these points have been growing for some time. So many columns have been written and so much airtime spent on how the world has spiralled in recent years – polarisation in the US, stagnation in Europe, Brexit, the rise of authoritarianism, the failure of global governance. For me none of it can be explained without understanding the centrality of what happened in 2008. Real wages and living standards collapsed. Public services deteriorated as governments scrambled to mitigate the damage to their public finances. The governments, treasuries and regulators that were supposed to curb the risky behaviour of banks and other financial institutions failed spectacularly. Neither the irresponsible spivs and speculators in the financial world, nor the technocrats who were supposed to rein them in, were properly held to account.

The result was fear, anger and insecurity for millions of people. For younger people in particular, their early years in the workforce are often characterised by a merry-go-round of insecure low-paid jobs, high rents, mounting debt, an inability to save for their futures, and house prices that remain wildly out of reach. As I discussed earlier, the financial crash damaged the central promise of capitalism – that your kids can do better than you, that reward is linked to how hard you work, that if you contribute, society will look after you. And that made millions of people very angry. Understandably so. We're still dealing with the politics of that anger today.

It is little wonder that, in this atmosphere of resentment and insecurity, people sought refuge in the politics of certainty. Populists who offer easy answers reassure people that they aren't alone in their grievances, that there are enemies to be blamed, that the elites who had benefited from the old status quo should have their rulebooks torn up and their institutions torn down. There has always been something comforting about retreating into our tribes and reducing our troubles to 'us versus them'.

The Tony Blair Institute for Global Change defines cultural populism as the claim that 'the true people are the native members of the nation-state, and outsiders can include immigrants, criminals, ethnic and religious minorities, and cosmopolitan elites'. Its research claims that the number of populists in power around the world increased a remarkable five-fold, from four to twenty, between 1990 and 2018.

Cultural populism emphasises religious traditionalism, law and order, sovereignty, and painting migrants as enemies. The symptoms are evident in the 'culture war' debates around the world: gun rights or the teaching of 'critical race theory' in the US; attempts to make Catalan and not Spanish the language of the classroom in Catalunya; the clash between Indian secularism and the Hindu nationalist Hindutva agenda; deforestation of the Amazon and the rights of indigenous populations in rural Brazil; or the rise of anti-vax movements just about everywhere during the Covid-19 pandemic. In the UK during the 2010s, cultural populism found expression in the resurgence of Scottish nationalism, the leftist populism of the Labour Party under leader Jeremy Corbyn, and the anti-immigrant and pro-NHS-spending (rhetorically at least) 'Vote Leave' Brexit campaign. But the figurehead of this cultural populist wave has undoubtedly been Donald Trump, with his America First agenda and promise to Make America Great Again. Cultural populism is also at the core of Vladimir Putin's Russian pseudo-imperialism, Viktor Orbán's Hungarian nationalism, the Hindu nationalism of India's Narendra Modi, Brazil's 'Trump of the Tropics' Jair Bolsonaro, Giorgia Meloni's Fratelli d'Italia and countless other insurgent nationalist movements.

While the policy agendas of these leaders and movements can vary starkly, they share common themes: rejection of corrosive 'others', whether they are powerful figures like bankers or subordinate ones like immigrant communities; nostalgia for a romanticised version of their national past (the 'again' in Make America Great Again); resentment of foreign influence; and distrust of establishment institutions they claim have been captured by out-of-touch elites at the expense of honest working people. Cultural populists demand – in the words of the Brexiteers – to 'take back control'. They speak to the anger and resentment of those who feel left behind economically and socially.

Alongside the rise of cultural populism has been a resurgence of national-sovereignty-based politics – asserting stricter control over borders and who can and can't live and work in a country, protectionist economic policies aimed at boosting strategically important domestic industrial sectors, a reassertion of national primacy over international courts or conventions, and antagonism towards and within international bodies. A leader column in *The Economist* in 2024 went so far as to lament that 'the liberal international order is slowly coming apart' – and this was before President Trump's re-election:

> As we report, the disintegration of the old order is visible everywhere. Sanctions are used four times as much as they were during the 1990s; America has recently imposed 'secondary' penalties on entities that support Russia's armies. A subsidy war is under way, as countries seek to copy China's and America's vast state backing for green manufacturing. Although the dollar remains dominant and emerging economies are more resilient, global capital flows are starting to fragment [. . .] The institutions that safeguarded the old system are either already defunct or fast losing credibility.

It is not just larger-than-life left- and right-wing populists who are responsible for the reassertion of national sovereignty. Former President Biden's government did as much as any US administration in decades to privilege American industry at the expense of global free trade – as much as if not more than President Trump's first administration that preceded it. Under Biden, the US government offered hundreds of billions of dollars in subsidies for domestic green energy, electric cars and semiconductors, massively ramped up the scrutiny on foreign inward investment, and imposed export bans against China, particularly in the high-end chips and chipmaking equipment sector necessary for the construction of large-scale AI infrastructure. The latter is a major factor in the increasingly fraught geopolitical race for AI dominance, which we'll explore in a later chapter. The US government's aggressive 'Bidenomics' industrial policy set off what *The Economist* described as 'a dangerous spiral into protectionism worldwide':

> Build a chipmaking plant in India and the government will stump up half the cost; build one in South Korea and you can avail yourself of generous tax breaks. Should seven other market economies that have announced policies for 'strategic' sectors since 2020 match America's spending as a share of GDP, total outlays would reach $1.1trn. [In 2022] nearly a third of the cross-border business deals that came to the attention of European officials received detailed scrutiny. Countries with the raw materials needed to make batteries are eyeing export controls. Indonesia has banned nickel exports; Argentina, Bolivia and Chile may soon collaborate, OPEC-style, on the output of their lithium mines.

President Trump may have denounced Bidenomics as the alleged cause of inflation and high prices during the 2024 Presidential campaign, but his administration's promises of sweeping trade restrictions on allies and adversaries alike build on Biden-era economic protectionism, which in turn preserved much of the trade policy of Trump's first administration.

In the UK, under the leadership of an increasingly populist succession of Conservative leaders, Brexit provided the most dramatic and self-defeating example of political isolationism. But many on the left bought into Brexit too. Jeremy Corbyn had been a lifelong critic of the European Union and, despite his party's nominal support for Remain, and the pro-European leanings of many of the party's grassroots members and supporters, he couldn't bring himself to align with the anti-Brexit cause. His more centrist successor Keir Starmer has conspicuously refused to reverse Brexit, but has sought closer ties with the EU, particularly over defence issues in the wake of the Trump administration's reluctance to provide Ukraine with explicit security guarantees. Even the technocrats of the European Union, once the chief cheerleaders for economic integration and legal harmonisation, have increasingly turned their backs on free trade. Talk of European autonomy and sovereignty now dominates discourse in Brussels. The European Centre for International Political Economy think tank (ECIPE) even blames what it calls 'the Brussels effect' for bolstering the spread of protectionism worldwide, pointing to restrictive trade policies and the imposition of fragmented laws across sectors like transport and logistics, telecoms and digital services.

To be referred to as a 'globalist' is now an insult. And there are no better poster children for globalism than big internet platforms, especially social media platforms like Meta that used to have a stated mission to 'connect the world'. In a sense, companies like Meta are both the children of globalisation and the arch propagators of it. They could never have existed were it not for the seamless flow of data across borders and its ability to carry information and ideas at lightning speed, above and beyond the grasp of democrats and tyrants alike. As a result, they grew into a new type of pan-global corporation that became deeply embedded in societies and economies on every continent and across the democracy-to-autocracy spectrum. But the borderlessness that these platforms enable and embody is now increasingly seen as a problem and not an asset.

The desire for governments to reassert national sovereignty over the internet, and in doing so cut Big Tech down to size, is in part driven by populist – and sometimes anti-American – instincts. But, as Benedict Evans explains, it's not quite as simple as that:

> Some of this is undoubtedly nationalism and protectionism [. . .] But the core of it, I'd suggest, is the rather basic Westphalian principle that a country's government has the right to say what can happen in that country. This isn't just about 'China' versus 'the west' – different liberal democracies have different views on how free speech works, for example, and no one outside the USA cares or even knows what the US constitution says about it [. . .] The same variance applies to privacy, competition itself and a whole bunch of other issues, right down to very micro issues like whether an Uber driver is legally an employee, or Airbnb's impact on house prices.

For as long as I can remember, I have been an instinctive internationalist. I believe that openness between nations – whether it's the free flow of commerce, the sharing of scientific expertise or the open exchange of ideas and culture – is the surest path to peace and prosperity and the greatest bulwark against conflict. There are undoubtedly problems that we only stand a chance of solving if nations work together – from climate change to cross-border crime – and I believe that in order to rise to these challenges it is not only necessary but desirable for nations to pool a degree of their sovereignty for the global good. But while holding these views may make me a deeply unfashionable globalist, I'm also a passionate democrat. The will of the people matters. Governments should be responsive to the needs, desires, norms and sensitivities of their citizens. Democracies have a right and a responsibility to assert control over issues of national importance. Part of the reaction against globalisation is a reassertion of these principles. It is in this climate of anti-tech sentiment and deglobalisation that the new rules of the internet are being written. But these forces are expressing themselves in different ways in different places – and the lack of a coherent and cohesive approach poses a profound threat to the open internet.

While myriad laws have been passed or proposed around the world in the last five years, there are three democracies that have a disproportionate influence on the global policy debate; three giant regulatory planets exerting a gravitational pull over the future of the internet: the United States, the European Union and India. So how are the Americans, Europeans and Indians regulating the internet, and what does it mean for its future?

America: fighting with itself

The open, accessible and global internet we use today has been shaped by American companies and American values like free expression, transparency, accountability and the encouragement of innovation and entrepreneurship. But when it comes to domestic internet laws, the United States has been conspicuous in its inaction at a national level. Despite its historic technological leadership, America hasn't passed a major piece of federal internet legislation since 1996, when Section 230 of the Communications Decency Act was created to address liability for online content. The statute is designed to protect free expression by giving online services immunity from civil liability for the actions of their users while providing protections for platforms to moderate content.

Compromise and bipartisanship have fallen by the wayside in Washington DC during the Trump and Biden years. The House and the Senate have become playgrounds for name-calling and finger-pointing, where each side is determined to dig its heels in so as to frustrate its opponent, almost no matter what policy is being debated. Both sides have railed against the evils of Big Tech but – attempts to ban TikTok over fears of Chinese influence aside – they have found it very hard to decide what exactly should be done. Supposedly bipartisan bills to strengthen antitrust rules – like the American Innovation and Choice Online

Act proposed by Senators Amy Klobuchar and Chuck Grassley, or the Digital Consumer Protection Commission Act proposed by Elizabeth Warren and Lindsey Graham – have hit the buffers time and time again.

Social media is often accused of diametrically opposed things by the rival Republicans and Democrats, with – as we have discussed – critics on the right accusing companies like Meta of censoring conservative voices, and critics on the left lambasting them for allowing those same voices to spread hate, misinformation and conspiracy theories. But their highly charged tug of war – particularly over issues of free speech and censorship – has made agreement on issues like privacy, content regulation and updating Section 230 borderline impossible.

This partisan gridlock has funnelled US action on tech companies into two arenas: federal agencies like the Federal Trade Commission (FTC) on the one hand, where the focus falls on the powers at their disposal under antitrust law; and the individual states on the other. Under the leadership of Lina Khan, who led the FTC during the Biden administration, the commission was aggressive in taking on what it considers to be monopolistic practices across a range of industries. The youngest chair in the FTC's century-plus history, Khan was appointed by then-President Biden to shake things up, having risen to prominence in the antitrust space for her theory that anti-monopoly actions had been too focused on the narrow issue of consumer welfare – that is, whether or not monopolistic practices hurt consumers directly by artificially inflating prices – and not focused broadly enough on issues of market power. In 2021, the year Khan was appointed, the FTC issued forty-two letters of investigation over mergers and acquisitions – the highest in more than a decade and nearly twice the number issued the previous year. And Big Tech was firmly in her sights. The FTC has been active in bringing cases against Big Tech companies, accusing them of anti-competitive practices and harming consumers, and challenging their acquisitions. At various points, Google, Microsoft, Apple, Meta, Twitter, Nvidia and others have been in the firing line. But the commission has chalked up a lot of losses in that time, leading critics to question if Khan's broad and hyper-active approach was misjudged. At the time of writing, it is still early days for Khan's successor, Andrew Ferguson, but it seems likely his approach to Big Tech will mark a shift away from Khan's antitrust theories and focus instead on issues of censorship. When his appointment was announced in December 2024, President Trump praised Ferguson's 'proven record of standing up to Big Tech censorship, and protecting Freedom of Speech in our Great Country', and promised he will be 'the most America First, and pro-innovation FTC Chair in our Country's History'.

The gridlock on tech legislation in DC has also pushed action into the hands of the states. With no sign of privacy legislation coming at a federal level, California passed a Consumer Privacy Act inspired in part by the European Union's landmark General Data Protection Regulation (GDPR) law. Texas and Florida passed laws intended to stop social media companies from moderating content that, in the view of many Republicans, leads to the censoring of conservative voices (at the time of writing, these laws remain on hold after being challenged on First Amendment grounds in the Supreme Court). A number of states have passed a patchwork of local laws to curb social media use among minors, including Arkansas, Utah, Texas, California and Louisiana. And forty-one state attorneys-general got together to sue Meta over concerns that Instagram is harmful to the mental health of teenagers.

Whatever merits or flaws there may be in these various laws and court actions, it is striking that the nation that created the internet has been forced, through political inertia, to regulate it in such a piecemeal and disconnected way. The risk of this approach is that it results in a fracturing of the internet within America's own borders. Having once championed the open internet, the US is slowly carving it up internally with a patchwork of disparate rules across different states.

But there are plenty of areas where, theoretically at least, it is possible to imagine legislation that can be passed with bipartisan support. The US is crying out for a federal privacy law. It could do so much more to protect against foreign influence in elections. Section 230 could be updated to ensure platforms are only granted continued protection from liability for the content they carry if they can demonstrate that they have robust practices for identifying illegal content and quickly removing it. Congress could bring more transparency, accountability and oversight to the processes by which large internet companies make and enforce rules about what users can do or say on their services. It could set clear rules on data portability to make it easier for people to vote with their feet by moving their data between services, or rules to govern how platforms should share data for the public good. The US could create a new digital regulator to navigate the competing trade-offs in the digital space and join the dots between issues like content, data and economic impact – much like the way the Federal Communications Commission has successfully exercised regulatory oversight over telecoms and media. There's no logical (as opposed to political) reason why ideas like these should be gummed up by partisan discord. And yet, for all the sound and fury aimed at Big Tech, this sort of coordinated bipartisan action has proved a step too far for America's gridlocked political system. While the US has struggled to legislate at home, in the second Trump term it has been aggressive in pushing back on laws in other countries that it sees as little more than vehicles for imposing censorship on US companies or attempts to extract cash from them. And, in its eyes, by far the biggest offender is the European Union.

Europe: squandering its advantages

While the US has been slow to legislate, Europe has been hyperactive. The European Commission is in many ways the opposite of the US Congress – while the latter is designed to be a check and balance on regulatory action, Europe is custom built to write laws. As a result, the European Union prides itself on being a pioneer of tech regulation, creating the blueprints that (it believes) will be copied around the world. The last few years of policymaking in Europe mean the EU now has the most comprehensive regulatory rulebook in the world. GDPR was the first serious attempt anywhere to create a set of principles and rules around personal data in a way that is practical in the digital age, establishing that it is accessible to all, that it is transparent and accountable, and that you own your own data. The EU has since passed three huge pieces of tech regulation – the bumper Digital Markets Act (DMA), the Digital Services Act (DSA) and the AI Act.

The DMA is a vast and complex law aimed at directly regulating the biggest technology companies, those it deems to be 'gatekeepers' to the digital economy, and in doing so theoretically at least ensuring their market power doesn't prevent new entrants and choke off competition. It sets a high bar for qualifying for gatekeeper status, and then sets significant restrictions to prohibit those that do qualify from unfairly hindering competition, such as 'self-preferencing' by favouring their own products and services over those of their competitors or leveraging their market power to cut out potential competitors. And it comes with a big stick to ensure the gatekeepers comply, including fines of up to 10 per cent of their global turnover – something the Trump administration views as the equivalent of a tariff on US companies.

The DSA, on the other hand, is about what happens when you log on. It creates responsibilities and obligations for internet companies to tackle illegal content, harmful behaviour and other risks. And it does so primarily by requiring companies to be transparent about things like how they moderate content and the action they take against illegal content and behaviour, and by requiring these companies to give users more control over their experiences online, for example by giving them ways to manage their privacy settings, control the content they see, and access their own data.

The AI Act is fundamentally about defining high-risk AI

systems – those it considers pose a significant risk to people's rights and safety – and imposing strict requirements and oversight to mitigate potential harms. So, for example, it prohibits AI systems that are designed to manipulate individuals' behaviour in a way that is harmful or deceptive.

All three are complex pieces of legislation developed at a time when the relevant technology has been evolving rapidly. As a result, they're far from perfect – in some cases onerously restrictive, in others unhelpfully vague, confused or contradictory. Even the policymakers themselves seem to be divided at times over what these laws mean, with political leaders offering reassurances to tech leaders over what needs to be done to comply with the laws, only for the civil servants operating beneath them to interpret the letter of the law in a way which sometimes collides with product reality. But whatever the pros or cons of the laws themselves, it is striking how much the EU has been able to get done in comparison with the partisan bickering in the US. In terms of technological innovation, the US is the hare to the EU's tortoise. But in the race to set the rules of the internet, the tortoise has long since passed the hare.

Has this flurry of regulatory activity actually helped Europe? It certainly gives European citizens the most extensive set of digital rights and protections of any in the world. But whether it helps them economically is far from certain. Throughout the development of these laws, the European debate has been characterised by two competing and at times contradictory impulses: a tremendous appetite to cut US Big Tech companies down to size; and the desire to boost Europe's own tech competitiveness. Many elements of the DMA in particular have been designed to target the business practices of specific companies – for example, Apple's closed ecosystem of hardware, operating system and app stores, or Meta's targeted personalised advertising – rather than enable the growth of European tech start-ups. The principle behind the push for greater transparency and accountability that motivates much of the EU legislation is difficult to argue against. It's wise to seek to hold large platforms to account through things like greater data reporting and auditing of systems, rather than micromanaging decisions on individual pieces of content. It's right that accountability should have teeth and there should be sanctions. And it's reasonable to impose different and stricter obligations on larger platforms than smaller start-ups.

Even so, these complex and multifaceted regulations have not served Europe's own economic ambitions well. They have created an inconsistent and unpredictable environment for companies to operate in. One consequence of that uncertainty has been a reluctance on the part of global companies to make AI models or products available to European consumers and businesses – Meta, Google and others delayed rolling out their AI assistants in the EU, for example, and even European success stories like Volkswagen are increasingly leaving to build and launch their AI products in the US.

The fact is, Europe has a bigger problem. The biggest hindrance to the next tech giant emerging on European soil is Europe itself. Three decades ago, Europe represented about a quarter of global GDP. In 2022, it had dropped below 15 per cent. European companies are growing more slowly, generating poorer returns on investment, and investing less in research and development than their US counterparts. GDP per capita in the EU is half of that in the States, at approximately $40,000 per European compared to $80,000 per American. None of the top ten global companies are European. None of the dozen most valuable unicorns – start-ups valued at $1 billion or more – are European. Most of the world's AI foundation models originate from the United States (109), followed by China (twenty). The EU has just seven.

European leaders have stated repeatedly that one of their main goals is for Europe to compete with the US and China in tech. They long for the next Silicon Valley to emerge on mainland Europe. I share that ambition. As a proud European, I would love to see the next Meta, Alibaba or Google emerge on the continent. And Europe has all the necessary ingredients:

top-quality talent, a large domestic consumer market, deep and liquid capital markets, a climate of entrepreneurialism, top universities doing cutting-edge research, and a culture of experimentation and innovation. But it has been held back by its own failure, among other things, to properly complete a project that has been on the agenda since I was working in the European Commission in the 1990s: the digital single market.

I have travelled a lot in my Silicon Valley role, and when I visit different cities I often meet local entrepreneurs who use Meta's products in one form or another – from small businesses using Facebook, Instagram and WhatsApp to reach customers, to developers and creators who build features and experiences for augmented reality. In September 2023, for example, during a visit to Athens to meet the Greek prime minister and other ministers, I spent some time with a group of the country's leading start-up entrepreneurs. What was particularly striking in that conversation was that the participants – all of them hard-working tech entrepreneurs trying to seek out success in neighbouring markets – shared a strong concern about their ability to access other EU markets. They argued that scaling a company from Athens is hard because of the impossibly complex patchwork of different regulations in each EU Member State.

In September 2024, former European Central Bank President Mario Draghi published a report on the future of European competitiveness which painted a damning picture of regulatory overload. 'The problem is not that Europe lacks ideas or ambition,' he wrote. 'We have many talented researchers and entrepreneurs filing patents. But innovation is blocked at the next stage: we are failing to translate innovation into commercialisation, and innovative companies that want to scale up in Europe are hindered at every stage by inconsistent and restrictive regulations.'

The idea that Europe's single market is incomplete has been met too often with defensive incredulity by Brussels officials – not least the former Commissioner for the Internal Market, Thierry Breton, who literally banged his fist on the table in red-faced fury when I pointed out what seemed obvious to me during a private meeting in Brussels. Yet, for all the EU's market-wide regulatory activism, a small tech start-up like those in Athens, or anywhere else in Europe, still has to navigate twenty-seven different intellectual property laws, various rules for the licensing of content, and obstacles to the delivery of goods bought online in order to make its services available across the EU. Each Member State also has its own Data Protection Authority (DPA), the regulatory bodies with powers to investigate and act against companies they find to have breached data protection laws, which inevitably leads to inconsistent interpretations of law, mixed messages to companies on how to comply, and a general sense of unpredictability across the continent. Germany even has a DPA for each of its sixteen federal states. It's a mess.

The existential nature of this economic underperformance is thankfully increasingly recognised by EU leaders. In April 2024, Emmanuel Macron gave a speech at the Sorbonne in which he said starkly: 'Our Europe is mortal; it can die.' He warned that Europe was 'falling behind' and called for, among other things, the deepening of the single market and an active EU industrial policy to support research and development in areas of strategic economic importance, not least AI. Mark Zuckerberg and I met him four days later in a private meeting to discuss the prospects for AI development in France and Europe. While the Elysée Palace cognac was excellent, the tenor of the discussion was downbeat: Europe was falling rapidly behind the US and China and it wasn't obvious what could be done. To be fair, though, the president's faith in AI can hardly have been boosted when the AI-powered Ray-Ban Meta glasses Mark showed off mistook a Pierre Soulages painting for a solar panel.

Europe's internal market gives it both an inherent disadvantage and an inherent advantage compared to the US and China: the US and Chinese markets both have a single language, whereas the EU has to balance the needs of multiple countries of varying sizes, languages, cultures and national agendas. That disadvantage may never be fully overcome, but it can be mitigated if the political will exists to make the digital single market a reality. And if it can, then the EU can reap the benefit of its major advantage: the diversity of a continent of relatively well-educated and affluent consumers of very considerable size (even after Brexit).

The greatest boost to European tech will come not from new regulation, but from the realisation of a genuinely borderless market of hundreds of millions of digital consumers. While there are other key ingredients to Europe's future success – more flexible labour markets so companies can adapt quickly to changing market conditions, an integrated capital market to provide deeper financing to the private sector, a greater appetite for entrepreneurial risk – completing the digital single market is the single biggest endeavour Europe can undertake to ensure its competitiveness in the digital sphere, as well as the surest way for it to ensure that European values are at the heart of internet regulation around the world.

India: a rising power

The third planet is India. For many in the West, where so much of the discussion around technology is focused on American companies and debates between American and European elites, it is easy to underestimate the significance of India's role in the tech landscape. India is the world's largest democracy – home to more than 1.4 billion people, more than four times the population of the United States, of whom more than 900 million were eligible to vote in what was the largest ever election in world history in 2024. And it is the fastest-growing national market for technology products. More people use WhatsApp in India than anywhere else on the planet. As of January 2023, there were 692 million internet users in India, up from 624 million in 2021. This represents a digital adoption rate of around 48.7 per cent of the population – so it has far more headroom for growth than the more technologically mature markets in North America and Europe. India is expected to be the largest smartphone market in the world over the next decade.

India has been at the forefront of the global technology revolution for decades, a hotbed of engineering talent and entrepreneurialism. But for a long time it has primarily been an exporter of talent, with the best and brightest Indian engineers finding happy homes in Silicon Valley and elsewhere. That's why one of the main focuses of Indian tech strategy for the last decade or so has been about retaining Indian talent and building up its domestic industry. And with some success: since about 2009, Indian tech companies have built billion-dollar companies riding the wave of success of digital public infrastructure (DPI), otherwise known as the 'India Stack' – a public–private partnership that provides citizens with digital access for various uses, from 'welfare payments to loan applications'. DPI then led to the Unified Payment Interface (UPI), a seamless digital payment system which helped pioneer direct benefits transfer.

India has become a thriving market for homegrown start-ups, with more than seventy unicorns. India's ambition is to rapidly become a global powerhouse of AI. A large majority of Indians (71 per cent) feel positive about AI, making it one of the most enthusiastic populations in the world, after China and Saudi Arabia. India has the second-largest developer base and third-largest start-up ecosystem in the world. The government believes AI could add nearly a trillion dollars to the Indian economy by 2035.

For more than a decade, Indian politics has been dominated by one man – Prime Minister Narendra Modi – and his Bharatiya Janata Party (BJP). The BJP's specific brand of cultural populism is based on reasserting a specifically Hindu version of Indian national identity that minimises the nation's Muslim heritage and rejects its British colonial past, alongside a protectionist 'Make in India' domestic industrial agenda, and a focus on the building of India's reputation as a major player on the world stage. The BJP's time in office has been marked by steady economic growth, periods of ethno-religious unrest, and a hawkish approach to national security, including an aggressive stance against its majority-Muslim neighbour Pakistan and an ongoing border tussle with China which has virtually halted diplomatic engagement with Beijing. It has also been particularly strident in issuing takedown requests to tech companies for content critical of the government – and not just to social media companies like Meta and Twitter/X, the latter of which complied with, but criticised, the government's request to remove posts expressing support for farmers in northern India demanding higher prices for their crops. When Google's Gemini AI model generated answers stating that some experts had characterised Modi's policies as 'fascist', the government responded quickly – via X – by accusing Google of violating Indian law.

Like the European Union, India has been a hyperactive tech legislator in recent years. It too has passed bumper legislation to regulate digital markets and e-commerce as well as the telecoms sector. India's Personal Data Protection Bill regulates the collection, processing and storage of personal data by individuals, companies and the government. And the Modi government has expended significant political capital on modernising and incentivising India's tech sector, which it sees as a vital component of India's economic strategy and a source of significant political clout on the world stage.

Harnessing AI is a crucial part of Modi's 'Techade' digital strategy. At Meta, I saw at first hand the huge demand from Indian researchers, start-ups and developers for our open-source Llama models. For example, Llama has been used by fintech platform Yubi to create its own open-source language model called YubiBERT, which operates in thirteen Indian languages; researchers from AI4Bharat created IndicTrans, a neural machine translation model; and the Indian Institute of Technology (IIT) in Madras is developing a speech-to-text model using Llama to enable speech-to-text translations in regional Indian languages.

India's attitude to tech policy reveals a stark geostrategic tension. India is at a crossroads – caught between the open values it takes pride in as the world's largest democracy and that bolster its credibility as a player on the global stage, and the desire to emulate the top-down state control over the flow of information exemplified by its neighbour and rival China. So it is simultaneously speaking the language of openness to encourage outside investment in its domestic tech sector, while legislating to assert greater control over how information circulates, particularly when it is in the hands of US and other non-Indian platforms. Which path India ultimately takes will have profound consequences for the future of the internet, which I will come to in the next chapter.

Unstoppable force meets immovable object

John Perry Barlow's idealistic proclamation that governments of the world should leave cyberspace alone because 'You have no sovereignty where we gather' has turned out to be somewhat naive. Governments have a right to set the rules that govern society within their borders, and with the internet becoming so deeply enmeshed in their economies and the lives of their citizens, they were always going to find a way to impose their will on it eventually. But the fact that this is happening at this precise moment in history, as the politics of deglobalisation, cultural populism and national sovereignty are ascendant, has resulted in a wide variety of laws, all driven by similar motivations and political instincts, but ultimately amounting to a mish-mash of different policy approaches without much, if any, attempt to coordinate between jurisdictions.

The borderless global internet has come face to face with the realpolitik of hyper-national politics – an unstoppable force meeting an immovable object. The result is that the internet is starting to fragment. Even the Trump administration's attempts to liberate American companies from the assertion of digital sovereignty by Europe and others – which it views as attempts to impose censorship and extract cash from US companies – will likely do little to prevent this fragmentation. Indeed, the damage this antagonistic approach does to the transatlantic relationship could well accelerate it.

While the idealists of the early internet will no doubt lament the end of the utopian dream of a unified space beyond the clammy grip of petty politics, much of what's happening now was inevitable. The internet has matured and become part of our lives. New rules to govern how it operates are necessary. Some of the laws being passed are good, others bad, while many are well intentioned but flawed. All are attempts to wrestle with dilemmas of the digital age. The open internet, after years of massive and often unchecked growth and innovation, is imperfect and in need of guardrails. But we need to be careful not to throw the baby out with the bathwater, with laws based on instincts and not evidence. We need democracies to work together to harmonise their rules as much as possible. And we need to be careful not to sleepwalk into the most disastrous scenario of all. If we want to protect the digital economy and the democratic values at the heart of the open internet, we need to counteract the threat posed by the rise of an alternative internet model based on very different values – authoritarian state control, censorship and surveillance. To do that, we first need to understand how that alternative's spread is aided and abetted by the disassembling of the open internet.

CHAPTER 8

The End of the Internet's Golden Age

In late 2023, the United States government performed a little-noticed – but profoundly consequential – policy U-turn. In a statement by the Biden administration's trade representative Katherine Tai during negotiations at the World Trade Organization (WTO), the US effectively ended its long-standing support for open cross-border data flows. Specifically, Ambassador Tai formally withdrew US support for provisions in new global e-commerce rules prohibiting what's known in policy circles as 'data localisation' – where governments seek to limit or block the flow of data in and out of their countries and ensure data is stored locally. The US said it did this in order to make it easier for Congress to pass domestic tech regulations, with officials reportedly stating it was about 'balancing the right to regulate in the public interest and the need to address anticompetitive behavior in the digital economy'. Once again, national sovereignty collided with the open internet.

The US has not definitively abandoned its championing of open data flows. Many in the Biden administration were vocal supporters – indeed, many were caught off guard by the apparent U-turn – and it remains to be seen whether the Trump administration will promote data flows or use them as a bargaining chip in trade talks. The debate is far from settled. The

US position in the WTO negotiations was better understood as a posture – reserving the right to withdraw its support for open data flows if it deems it necessary in order to get domestic legislation on the books. But this is far from the only point of pressure on America's long-standing support for open data flows.

Arguably the most prominent US critic of open data flows is one-time Democrat Presidential hopeful Elizabeth Warren, who argues that Big Tech companies leverage trade deals to guarantee their ability to transfer data regardless of the privacy protections guaranteed in foreign jurisdictions. In effect, Warren argues, Big Tech is using open data flow provisions in trade agreements as a means to circumvent privacy obligations under their countries' domestic laws.

This perspective is flawed for several reasons. First, the US government has supported the free flow of information across borders for more than three decades – well before most of the tech companies that are being criticised for co-opting trade negotiations to their own ends even existed. Second, trade agreements don't constrain the US government's ability to regulate the activities of US companies; they simply ensure that there cannot be arbitrary or unjustifiable discrimination against another country's companies. And third, data protection laws usually attach corresponding rights to the data subject, which means that people can enforce their rights under their domestic laws wherever in the world the data is stored. In other words, if your concern is how to ensure that Big Tech complies with privacy laws and is brought to task for privacy violations, then restricting data flows is a complete red herring.

But Warren is not alone. Following the WTO U-turn, campaign groups like the European Consumer Organisation (BEUC), Access Now and Privacy International said that governments worldwide should reassess their digital trade policy to follow in the footsteps of the United States. Indeed, in the early 2020s we came perilously close to a metaphorical digital wall being built in the Atlantic. The EU has strict data protection regulations, primarily governed by GDPR, which restrict the transfer of personal data only to countries with adequate data protection standards. The United States, on the other hand, has different privacy laws, and some in Europe have expressed concerns about the level of protection afforded to personal data transferred to the US. In 2015, legal challenges by Austrian privacy activist Max Schrems led to the Court of Justice of the European Union (CJEU) invalidating the Safe Harbor framework, which had governed EU–US data transfers, citing concerns over US government surveillance practices revealed by Edward Snowden. This led to negotiations culminating in the establishment of the EU–US Privacy Shield framework in 2016. But this too was struck down in July 2020, when the CJEU invalidated the Privacy Shield agreement in the 'Schrems II' case, ruling that it did not provide adequate protection against US government surveillance. The court cited concerns over the lack of limitations on US surveillance activities and the absence of effective legal remedies for EU citizens.

This created an existential moment for the internet. Thousands of businesses on both sides of the pond – from huge corporations to small start-ups – rely on the transfer of data between the European Union and the United States, where most data centres by far are based. Following the invalidation of the Privacy Shield, companies have had to rely on other mechanisms, such as Standard Contractual Clauses (SCCs), to transfer data between the EU and the US. But without a blanket agreement between governments that would allow data to be transferred freely, many US companies – Meta included – had to consider the previously unthinkable: withdrawing their services from Europe.

Let's just think about that for a moment. For more than two decades, Europeans have grown accustomed to being able to search, share, buy, sell and surf using American-based platforms, apps and websites. But all of that relies on a legal arrangement between the US and the EU – and that arrangement came perilously close to collapse. If a new one couldn't be agreed, Europeans' experience of the internet would change dramatically, and an axe would have been taken to the European and transatlantic digital economy, and by extension the global economy too.

The EU and the US engaged in negotiations over a new data-transfer framework that addressed the concerns raised by the CJEU. This included discussions on enhancing privacy safeguards, strengthening oversight mechanisms and ensuring redress mechanisms for EU citizens. But progress was slow and hesitant, and there were times when it seemed as if a mutually satisfactory agreement might be out of reach. Eventually, however, in the summer of 2023 it was announced that a new Data Privacy Framework had been agreed.

What's really at stake in all this is how the internet works. For as long as data has flowed freely through the pipes of the open internet, innovation and commerce have boomed and free expression has flourished. We all benefit, as individuals and as societies. But America benefits most. Not only are its companies pre-eminent in the digital economy, but the soft power from the spread of its values and cultural norms is beyond measure. When the global internet breaks down into its component parts, these parts are, in turn, easier to control. Governments as well as private corporations will find it easier to impose their own rules on their localised network and will be able to disrupt – and eventually completely up-end – the freedom of expression that the internet currently enables. As they do, America's soft power will slowly but surely dissipate.

Perhaps curtailing American hegemony appeals to many people on some level. You can be sure that it appeals to America's most powerful critics, such as the Russian and Chinese governments. But let's now see what this actually means for the open internet and, by extension, for all of us who use and rely on it.

Data isn't oil

According to a Reuters report, Democratic Senator Ron Wyden, the chair of the Senate Finance Committee, called the US's apparent change of heart on data flows 'a win for China, plain and simple', arguing that it would strengthen the Chinese model of internet censorship and government surveillance. Others, like Jennifer Brody and Allie Funk of the Freedom House think tank, warned it 'risks further fragmenting the global internet, emboldening authoritarian governments and their aspiring counterparts, and violating rights around the world. Particularly for people living in countries that already have data localization requirements, the impact on human rights is grave.'

Indeed, as governments around the world seek to exert ever greater sovereignty over the internet, data localisation has proved to be a seductive idea for policymakers. But while the US was still using its clout on the international stage to preserve open data flows and resist localisation, the risk that the infrastructure of the internet would fracture remained largely contained.

While Chinese-style 'hard' data-localisation policies result in an almost complete enclosure of a country's data economy within national boundaries, a 'soft' variation has gained traction in many open democracies. This milder form requires data to be mirrored in local servers, ensuring copies are held domestically, which can slow internet services and limit access to them. Different styles of data localisation have found support around the world, in both democracies and more autocratic states – from India and Indonesia to Turkey, Vietnam and a number of African states.

When liberal democracies are the ones championing data-localisation policies, they unwittingly give legitimacy to authoritarian governments who want more direct control over the internet. Governments around the world are now growing ever more aggressive in their demands for platforms to comply with rules to produce data, block content and break the strict end-to-end encryption that keeps messaging services private and secure. In the words of the Center for Strategic and International Studies: 'National security justifications for these mandates are often thinly veiled attempts at asserting greater control of the domestic digital domain; meanwhile, data localization has had negative impacts on human rights, privacy, and economic interests.'

Cultural populists often argue that you can have both greater national sovereignty and a stronger economy. Boris Johnson, the bumbling British prime minister who led the Brexit campaign, argued that Britain could 'have its cake and eat it' upon leaving the European Union. Despite withdrawing from the world's largest economic single market, he claimed his Brexit deal meant the UK could 'go our own way but also have free trade' with the EU. Needless to say, this was deeply disingenuous, and much of it has been shown to be wrong. Brexit has hurt Britain's economy, taking a chunk out of its GDP, depressing trade and investment, and inflating food prices, both exacerbating the impact of the Covid-19 pandemic and hindering the UK's ability to recover from it. Much-vaunted trade deals, particularly with the US, have failed to materialise. Sectors like transport, hospitality and retail have been hit by labour shortages.

Perhaps just as important, however, if more intangible, is the hit from Brexit to Britain's prestige and global influence. From my vantage point on the west coast of America for several years, it is telling how little the UK features in the strategic considerations of Silicon Valley. Inside Meta, and in the US more generally, I was asked far more often about the travails of Prince Harry and Meghan Markle than about British politics or foreign policy. (Mark Zuckerberg didn't visit the UK once in my time with the company, and I saw little reason to encourage him to do so. We did, however, make the trip to mainland Europe on multiple occasions to meet leaders and regulators in Brussels, Paris and Berlin.)

Likewise, leaders from the EU's Thierry Breton to India's Narendra Modi have claimed that greater sovereignty over the data economy within their respective borders will boost their economic competitiveness. But protectionism is protectionism, whether it is on- or offline. You may want to have your cake and eat it, with both greater economic competitiveness and pro- tectionist policies, but you can't. Divorcing yourself from the global flow of goods and services may boost a specific industry for a while, but it will hurt your economy overall.

Data-localisation advocates often base their arguments on the idea that data is like oil. The argument goes that a nation's citizens produce this commodity, but its value is then taken by foreign – and specifically American – companies to the benefit of investors and employees beyond their borders. If a country wants the value to remain with it, so these advocates argue, it needs to keep the data within its borders. A by-product of doing so, happily for digital protectionists, is that if foreign Big Tech companies then want to access this data, they will have to pay up by building expensive local data centres and employing local citizens to run them and oversee their business in the country. Of course, controlling the infrastructure through which data flows also makes it easier to impose rules on what that data includes – in other words, about what is and isn't allowed online – which makes it particularly attractive to authoritarian regimes that want to censor content or keep their citizens under surveillance.

The problem with this data-is-oil analogy is that data isn't a scarce and finite resource to be owned, hoarded and then traded. Its value doesn't derive from being pumped from source to wherever it is going to be used up. On the contrary, its ability to circulate and flow across borders is fundamental to how it creates value. But to understand why restricting the flow of data is essentially inimical to its use and purpose, it's worth considering what we mean by data in the first place, and why it has become so valuable.

Data is information that has been ordered in such a way that it can be systematically interrogated and used for purposes that wouldn't otherwise have been possible. A narrative history book about medieval England will be full of fascinating information and revealing stories, but it isn't easily quantifiable data. On the other hand, the Domesday Book, which surveyed and valued landed property in late eleventh-century England, was the most comprehensive exercise in data collection of its day. Quantitative data has been fundamental to the development of the modern state. Modern epidemiology, for example, relies on the work of early data pioneers like William Farr, who created the first national vital statistics system in the UK in the middle of the nineteenth century.

The digital age fundamentally changed our ability to gather and utilise data, turbocharging its impact on society. Vellum parchment gave us the Domesday Book, but copying data was a labour-intensive task, so its impact was limited. Little by little, however, technology has allowed us to make more use of it. From index-card systems to IBM's automated machines, the invention of digital memory, microprocessors, the network, the internet, mobile technology and now AI – at every stage, new technology has massively increased the utility of data. In the age of AI models that process billions upon billions of data points, even the lengthy prose of that medieval England history text has become usable data.

It's hard to overstate how important digital services have become to today's global economy. Processing and analysing vast quantities of data – often with the use of sophisticated AI-based tools – is now woven into the fabric of how organisations operate across every sector and sphere of public life. This proliferation of data-driven tools has been a boon for the global economy, and it has also helped to democratise access to it, levelling the playing field between small businesses and big corporations. Just about anyone can now start a business online without the need for a big bank loan to pay for major overheads like renting a shop front or office space, thanks to e-commerce platforms, social media apps and easy-to-design custom websites. And with personalised digital ads, small or localised businesses can reach targeted audiences of potential customers for just a few pounds, euros or dollars, rather than needing the deep marketing budgets required for mass-market television, radio or billboard campaigns. This is especially true for those in rural communities and developing economies, where people with enormous talent and potential have long been held back by poor infrastructure and long distances from metropolitan economic centres.

It is not just the global economy that relies on data sharing across borders. Data sharing is crucial for national security, for example through global data-sharing initiatives aimed at clamping down on serious crime and terrorism, like the CLOUD Act and the Budapest Convention on Cyber Crime.

Crucially, data is a 'non-rivalrous good', which is a technical way of saying it doesn't get used up when it's consumed. Burn oil, and it's gone for ever. Make use of a data point, and it still exists to be used over and over again. Having a great big stockpile of it is not in itself particularly valuable. It's what you do with it that counts. The value of data stems from the quality of insights it can produce. Think of it this way: a beautifully crafted poem is more valuable than a long list of random words. The same principle applies to databases – a large database that includes random data points has little or no value (at least to organisations that don't have highly sophisticated systems capable of analysing it), while a small database with well-connected data points can be invaluable. Unlike oil, the value of data depends on the context within which it is placed, especially when it comes to the highly specific contexts required to deliver personalised services.

A database about people's clothing preferences is much more valuable to a clothing retailer than it is to a restaurant chain, and vice versa for a database full of people's dining preferences. Yet the source of the data in both cases could be the same individual with a profile in the H&M and OpenTable websites. And, unlike oil, the value of data reduces over time. That is, the value of this year's data is much greater than the value of last year's data, and so on.

Of course, a big cache of data can be extremely valuable for organisations who know how to make use of what they have – whether they are a Silicon Valley tech company, a German car manufacturer, a local fast-food delivery service, a charity or a government health ministry. But its value is very different to that of a scarce natural resource, and it is not constrained by how often or where it is processed, utilised and consumed.

No analogy is perfect, but a better liquid to liken data to is water, with the global internet like a great borderless ocean of currents and tides. A single drop may be bland and limited in value, but allowed to flow freely it becomes the key ingredient of the life of the internet. In its constant churning, mixing and endless recycling, complex ecosystems of innovation and commerce can emerge and thrive. To contain data – to fix it geographically and to restrict its flow to national borders – would be to turn this great ocean of innovation into a still lake. The global internet is built on this principle of cross-border data flows – just as the global economy relies on capital, people and technological innovation to cross borders in order to flourish.

The fracturing of the global internet

It would be a tragedy for the internet if the great democracies of the world were the ones which chose to block up its pipework. But perhaps the most profound threat to the open internet comes when nations decide to build a completely different network of pipes altogether. The open internet relies on common standards and protocols, overseen by international governance bodies, but there's nothing inevitable about this. If countries opt out of these arrangements, or are forced out, and begin developing and implementing their own standards, then we will be in a world where not only do governments wish to block or restrict the flow of data across borders, but also the infrastructure of the internet in one place is completely incompatible with that in another.

Recall the example in Chapter 3 of the Chinese version of the internet, with its great firewall all around it, sifting and blocking any data that enters or leaves. The Chinese government's purpose in building this firewall isn't simply to cut off its internet from the wider world. Rather, it wants to embed the very principle of its design into the architecture of the wider internet – or rather, internets. It has made moves to shift the development of internet standards away from the traditional collaborative, consensus-based forums of the IETF, IEEE or W3C and towards the United Nations International Telecommunication Union (ITU), where individual nation states have more direct sway, and where it has pushed its model of 'New IP', developed by Chinese networking giant Huawei. New IP (New Internet Protocol) is designed to build 'intrinsic security' into the internet, which 'means that individuals must register to use the internet, and authorities can shut off an individual user's internet access at any time', according to Mark Montgomery and Theo Lebryk in the NYU-based justsecurity.org forum. This approach completely up-ends the way the architecture of the internet is decided. As *Foreign Policy* magazine put it, 'China wants to run your internet':

> The battle for the internet governance of the future will differ from past struggles over technical standards in a fundamental way. Setting these rules is not exclusively about addressing technical issues or projecting global power. It is about promoting different visions of the world: a decentralized and democratic one (the traditional internet) or a centralized and authoritarian one (China's 'New IP'). This is an entirely new chapter in the history of standards setting that will contribute to shape the relationship between China and the West, with enormous geopolitical and economic ramifications.

Russia too is a key part of this, but for different reasons. Having made itself a pariah state with its invasion of Ukraine, it has withdrawn from international governance bodies like the Council of Europe and been suspended from the European Broadcasting Union. The Ukrainian government has also called for Russia's access to the internet's domain name system, overseen by ICANN, to be withdrawn, which would effectively remove .ru websites from the internet. This move was rejected by ICANN. '[T]he internet is a decentralized system. No one actor has the ability to control it or shut it down,' wrote its CEO Göran Marby in his response. 'Essentially, ICANN has been built to ensure that the Internet works, not for its coordination role to be used to stop it from working.' If countries get locked out of the internet's voluntary governance bodies, it is not hard to imagine them establishing their own rival ones. And once the internet's governance becomes splintered, it won't take long for its pipework to follow.

In such a scenario, the internet will no longer exist as a global communications entity. Entire rival internets will exist, different not just in the rules that govern how people can use them, but different technologically. These internets will diverge ever more starkly as technologies develop along completely different tracks. The 'American internet' won't simply look different to the 'Chinese internet' or the 'Russian internet'; they will be fundamentally different entities. The 'European internet' is already diverging from its US counterpart, and could do so more dramatically if the flow of data between the continents is restricted. The 'Indian internet' could follow suit. And countries all across the global South could find themselves increasingly technologically isolated, especially if they are locked out from accessing foundational AI technologies built in the US.

If we continue down the path to internet fragmentation, there won't be one great watershed moment when we wake up one day and find the internet has changed beyond recognition. The change won't come about as the result of a major shock – like a financial crash or the outbreak of war – that could rally the democratic world to act with urgency to restore it. Rather, like the proverbial frog in the pan of boiling water, we will probably barely notice the small, gradual changes, but by the time we do it will be too late. According to the Internet Society, the world we find ourselves in by that point might look a bit like this:

> Let's say that on a regular day you visit Facebook, Wikipedia, and Google. With an open Internet, you are able to visit these sites without giving it much thought. Now imagine this happening instead:

facebook.com – Blocked. You can't get there. en.wikipedia.org – You should get to English Wikipedia, but instead you see what looks like a version of Wikipedia in a different language. And you aren't even sure if it is *the real* Wikipedia. google.com – You wind up not on Google, but instead on some other search engine, searching only government approved sources. This is a splinternet. Where the addresses you normally use on the open Internet can take you to completely different places – or sites can be blocked entirely. It's where you can't trust the names and addresses to take you to where they're supposed to. It's where borders are added to a borderless system. It's where the free flow of information becomes restricted and suppressed [...] You may be able to use the same browsers and mail programs, but you can't get to the same places. And even if you can, you have no idea if the local government is monitoring everything you do.

Here are some examples of how we, as western citizens and consumers living in liberal democracies, might experience a fractured internet in practice:

- World news: getting live reports from the other side of the world will not be as fast or cheap as it is today. Some digital books and newspapers will also be restricted in some parts of the world, or they will simply be too expensive to buy.
- Streaming services: YouTube, Spotify, Netflix and other services could be vastly different and their digital content libraries emptier. And forget about accessing videos from some parts of the world – they will probably be very costly, unreliable or plain inaccessible.
- Social media: it will become slower, especially in smaller and less developed countries that might lack good data infrastructure. Information from your friends in certain countries may not be available.
- Email: some emails from certain regions of the world will never reach your inbox because the country they originated from blocked them.

- Small businesses: millions of businesses will lose access
  to the global digital market; they will only have access to
  regional or national digital markets. Advertising will become
  more costly for them, since processing data will also be more
  expensive for internet providers and platforms, who will
  probably pass on that cost to users.
- Banks: cross-border money transfers will be delayed or
  halted completely, depending on the security and cost of
  each country's digital networks.

The fragmentation of the internet will also make it harder for police authorities to detect criminal activity across the globe, while making it easier for some countries to provide safe havens for illegal digital activity. And it would almost certainly lead to a significant decline in international cooperation for academic research, which could have far-reaching consequences. Researchers will have to comply with increasingly stringent laws around data use and storage in each country. They will not be able to share and aggregate data as easily (or at all), resulting in an end to international research in some cases or at best slowing down those types of collaboration. For example, it could affect the development of car safety measures, as manufacturers aggregate data from cars across the world and are able to improve safety features like brakes or airbags. It could slow AI-based medical research: advances in cancer research and other areas of healthcare have been accelerated recently by the impact of AI – but AI is by default a technology that relies on aggregating data. And it could hinder consumer product research: A/B testing is a method that improves user interfaces of most digital products – but with the shrinking of aggregated international data, companies will not be able to improve their products as fast and efficiently as they do now. This will hinder the development of research in countries with less access to data, and favour companies that have access to large swathes of data, like those in the US and India.

We may soon look back wistfully at the early twenty-first

century with nostalgia for the internet's golden age. In the future, there could be dozens of national or regional internets, each with their own rules and underpinned by their own values and standards, potentially running on incompatible technologies. Your online experience could vary massively depending where in the world you are – and in many ways it will probably be far more limited than today's internet, no matter where you are. As the internet fragments, we in the West are losing something we have come to take for granted. But the fact that this is happening at a crucial time in the early stages of the AI revolution could be an omen of an even more starkly technologically divided world in the years to come.

CHAPTER 9

Superpowered Superpowers

Right now, the name of the game in AI is the building of ever more powerful 'foundational AI models' – the vast large language or multi-modal AI models that generative AI products are built on. Doing this requires a level of energy-intensive computing power that is currently the preserve of only a handful of private sector companies. The same companies' immense research and development spending exerts a gravitational pull on the world's computer-science talent, centralising expertise and institutional knowledge within the confines of their campuses and satellite offices. And the models they are building require state-of-the-art computer chips that are manufactured in very few places by very few suppliers (vaulting chipmaker Nvidia to a $3.2 trillion valuation as AI fever gripped the markets in 2024), which means demand is high and supply is short. It's little wonder then that the AI race has moved beyond the realm of private sector competition and into that of geopolitics.

There are two superpowers in the global AI race: the United States and China. Under President Xi Jinping, China has been transformed into a tech powerhouse. The Chinese government is actively and aggressively promoting tech sectors like electric vehicle production and quantum computing. And it is putting the weight of the state behind a major push for primacy in business-facing AI tools. For all China's efforts, though, right now the US appears still to be winning the AI race. It is home to Silicon Valley powerhouse companies who have the keys to the world's most advanced AI infrastructure and the deep pockets to kit it out with state-of-the-art computer chips. It is home to world-leading universities doing cutting-edge research. It attracts the top talent from around the world. And it has enormous economic and diplomatic clout on the world stage.

And yet, despite the US's AI primacy, it has the jitters. It knows its dominance, like so much else, is neither inevitable nor permanent. The threat China's economic and technological power poses to the United States has done more than any other factor to spook it into its lurch towards protectionism.

An important part of President Biden's industrial policy was trying to cut China out of the market for key elements of AI kit. The Biden administration issued two decisions through the Department of Commerce's Bureau of Industry and Security (BIS) in 2022 and 2023, as well as an executive order in 2023, aimed at cutting off Chinese access to vital technology. The first BIS order implemented export controls to restrict China's ability to 'both purchase and manufacture certain high-end chips used in military applications'. The second order closed loopholes that allowed Chinese companies to buy Nvidia chips, which the previous order did not adequately address. President Biden also said in a letter to Congress that he was declaring a national emergency to deal with the threat of advancement by countries like China 'in sensitive technologies and products critical to the military, intelligence, surveillance or cyber-enabled capabilities', and signed an executive order in August 2023 that authorised the US Treasury Secretary to prohibit or restrict US investments in Chinese entities in three sectors: semiconductors and microelectronics; quantum information technologies; and certain artificial intelligence systems. The order aimed to prevent American capital and expertise from helping China develop technologies that could support its military modernisation and undermine US national security.

Congress also passed the Creating Helpful Incentives to

Produce Semiconductors for America Act, or CHIPS Act, which was signed into law by President Biden in August 2022. The Act allocated $52.7 billion in federal subsidies for semiconductor manufacturing and research, but excludes what it refers to as 'foreign entities of concern' – including companies owned by, controlled by, or subject to the jurisdiction of China – from financial incentives for semiconductor manufacturing in the US, and prohibits recipients of CHIPS funding from expanding semiconductor manufacturing in China and other countries that pose a threat to US national security. The intent of these export controls is clear: the US wants to restrict China's AI ambitions by denying it access to the most advanced – and, crucially, US-designed – computer chip hardware. Or, as *Time* magazine put it, 'the US government committed to stopping China from becoming an AI-enabled authoritarian superpower'.

This is not an issue on which Democrats and Republicans disagree in any meaningful way. The Biden administration's export controls came on top of measures introduced by President Trump's administration in his first term, which placed tariffs on around two thirds of imports from China, and there is every sign that the second Trump administration intends to go much further still. This slow decoupling of the US and Chinese economies – or 'de-risking', as the Americans describe it – has been an active policy over many years. As the Chinese economy gradually opened up during the boom years of globalisation, the US and China became more economically integrated. China provided the factory floor, the US provided the goods-hungry consumers. Then came the financial crash in 2008, the elevation of Xi Jinping in 2012 and Trump in 2016, and the Covid-19 pandemic in 2020, the last of which threw global supply chains reliant on Chinese exports into disarray.

China has been centralising control over its technology sector under Xi, particularly since the creation of the Cyberspace Administration more than a decade ago. It responded to the US export controls with its own, including on the export of minerals such as germanium, gallium and graphite that are considered key elements of high-tech systems needed for green energy, defence and a host of other strategically important areas. China has also continued to develop its own AI hardware capabilities, in line with its 'Made in China 2025' strategy to bolster the country's domestic tech industry and increase its self-reliance. In 2023, Huawei released the Mate 60 Pro, which scholar Chris Miller described as 'the most "Chinese" advanced smartphone ever made'. It contained a new 7-nanometre chip that is supposedly comparable to Nvidia's A100 chip. The US may hold the lead in chip technology, but its hopes that strict export bans can keep Chinese companies from developing high-performing hardware of their own may be shortlived.

Chinese progress on generative AI is held back by something else – the Communist Party's determination to control public data and discourse. In 2023, the government proposed new rules insisting that 'content generated using generative AI shall embody the Core Socialist Values and must not incite subversion of national sovereignty or the overturn of the socialist system'. That means no criticism of the party's leadership or its policies, no anti-party sloganeering, and no political satire poking fun at the nation's leaders. The desire for political control limits innovation – at least in terms of developing systems to generate text, audio and video content for citizens to consume. Chinese LLMs, it is reported, are often trained on smaller and more restricted data sets, and they also present big challenges to engineers as they seek to ensure the systems don't spit out non-compliant answers.

That said, some now believe that some of the most advanced Chinese models, including the DeepSeek-R1 model released in early 2025 and Alibaba's Qwen models, are keeping up with or even surpassing the leading US models. Notably, many of these models are open-sourced, a clear sign of the intention to spread Chinese-based AI models around the world. And the Chinese government sees a particular opportunity in taking advantage of the vast stores of corporate data held by Chinese companies. As *The Economist* explains:

Corporate applications need corporate data, a lot of which is squirrelled away inside companies. So the other plank of

China's strategy is to turn corporate data into a public good. The state does not want to own the data but [...] to control the channels through which it flows. To that end, the government is promoting data exchanges. These are meant to let businesses trade information, packaged into standardised products, about all areas of commercial life, from activity at individual factories to sales data at individual shops. Small firms will gain access to knowledge once reserved for the tech giants. Banks and brokers will get a real-time picture of the economy.

The US still has the advantage on advanced AI models right now, but for how long is anyone's guess. When I have spoken to US officials and foreign-policy experts, they agree that an aggressive trade policy will delay the point at which China catches up, but it won't keep the wolf from the door for ever. And given the extent to which US and Chinese economic interests became entwined in the early twenty-first century, the more the trade war escalates, the bigger the economic risk the US is taking. In 2023, US Treasury Secretary Janet Yellen warned that 'a full separation of our economies would be disastrous for both countries'. And in a piece decrying what it called the 'destructive new logic that threatens globalisation', *The Economist* warned that economic conflict between the US and China looks increasingly inevitable:

Some [US politicians] simply want to stop China becoming too rich – as if impoverishing 1.4bn people were either moral or likely to ensure peace. Others, more wisely, focus on increasing America's economic resilience and maintaining its military edge. A reindustrialisation of the heartland, they argue, will rekindle support for market capitalism. In the meantime, as the global hegemon, America can weather other countries' complaints. This thinking is misguided. If zero-sum policies were seen as a success, abandoning them would only become harder. In reality, even if they do remake American industry, their overall effect is more likely to cause harm by corroding global security, holding back growth and raising the cost of the green transition.

The great AI divide

Where, then, does this leave everyone else? International trade law expert Anu Bradford has argued that the US–China trade war has prompted other countries to introduce more protectionist trade policies of their own. 'As a result,' she argues, 'the tech war risks entrenching techno-nationalism as a global norm. This can be seen as a victory for the Chinese state-driven model as governments are abandoning the US's vision of an open, free, and global digital economy.' Certainly, the race for dominance in foundational AI models poses a big dilemma for other nations: do they attempt to join the AI arms race themselves for fear of being left behind, or become supplicant states to the US or China? Attempting to develop their own vertical AI infrastructure – from AI software and applications to data centres and chips – would be a hugely expensive and energy-intensive drain on national resources. And with the entrenched advantages that the US and China already have – huge domestic tech sectors, aggressive national strategies and a stranglehold on much of the world's high-end engineering talent – how on earth do they catch up? As the local replies in the old Irish joke about the tourist asking for directions in Dublin, 'I wouldn't start from here if I were you.'

Even for larger economies, state-funded AI development programmes may struggle to compete with the resources available to the large US tech companies. Allocating public resources to building specialised data centres would also be highly inefficient, with dramatic opportunity costs (meaning the loss of potential gain from alternatives) as funding is diverted away from public services.

This may be an argument for governments stepping back from major investments in AI model development. And yet, in this age of deglobalisation it is hard to imagine any advanced economy choosing not to invest in a sovereign AI capacity of some kind, as their resulting dependence on US or Chinese technology could leave them at a profound strategic disadvantage over the long term.

That's not to say that some nations don't have distinct advantages of their own in the global AI race. Goldman Sachs's Jared Cohen and George Lee describe a number of what they term 'geopolitical swing states', which each have one or more of four key elements necessary to be big AI players: 'economic or regulatory power to shape global technology innovation and commercialization; differentiated AI technology and talent ecosystems; world-leading companies that give them control over critical chokepoints in the AI supply chain; clear national AI strategies and the capability and will to implement them and to deploy capital'. These 'swing states' are:

- The EU, with its leadership on regulation and huge market of 450 million citizens, and in particular France, which has a strong domestic talent pool and determined technological strategic leadership under Emmanuel Macron;
- The UK, which also has a big talent pool and world-leading research universities;
- The United Arab Emirates, whose ambitious state-led AI strategy is the driving force behind the development of its sophisticated Arabic-language open-source LLMs, including Falcon and Jais, and which is home to its own AI and cloud-computing companies, which have recruited talent from across the Middle East, Asia and beyond;
- Israel, which has been a hub for tech leadership in a number of fields for many years, particularly in defence and cybersecurity;
- India, with its fast-growing economy, enormous domestic market and ambitious national tech strategy; and
- Japan, the Netherlands, South Korea and Taiwan, all of which play a vital role in semiconductor manufacturing and supply chains.

But while these nations have significant cards to play, they don't come close to the hands that the US and China are holding. And while smaller and less developed countries may also try to compete, they simply do not have the economic clout to develop their own AI foundation models or data centres on the scale required. Many leaders, particularly in the global South, fear they could be left behind in a new era of AI haves and have-nots.

The International Monetary Fund has argued that AI will widen the margin of inequality between rich and poor nations by 'shifting more investment to advanced economies where automation is already established'. A similar argument has been made by the Bank for International Settlements and by MIT economist Daron Acemoğlu. And, as Pew Research suggests, a lot of experts believe that technological changes in the future, including AI, will have a mostly negative impact, including a rise in economic inequality.

For many of these nations, the best hope of passing on the benefits of the AI era to their own citizens and economies is through access to the foundational models built and operated elsewhere, essentially piggybacking on the computing power and energy resources of the big players in order to adapt models and develop an application layer tailored to the needs of their own societies. This takes us into one of the behind-the-scenes debates that has erupted in policy circles in the post-ChatGPT world. Should these huge AI models be made widely available through open sourcing in a way that allows the benefits to be spread widely around the world? Or should proprietary models be kept under lock and key by the big Silicon Valley companies, potentially on the orders of the US government, for fear that making them openly available will mean bad guys can use them to their own nefarious ends?

For a host of reasons, I believe firmly that openness is the way to go. The infrastructure of the internet runs on open-source code, as do web browsers and many of the apps that billions use every day. The cybersecurity industry exists because of it. Meta has a long history of sharing AI technologies like PyTorch, the leading machine-learning framework, and the various versions of Llama. We've seen at first hand how making AI models available to researchers can reap enormous benefits. Meta's models have been used to speed up the discovery of antibiotics, generate equally accurate and detailed MRIs using about a quarter of the raw data, and help preserve the world's language diversity.

Like all 'base layer' technologies – from radio transmitters to internet operating systems – there will be a multitude of uses for AI models, some predictable and some not. And like every technology, AI will be used for both good and bad ends by good and bad people. The response to that uncertainty cannot simply rest on the hope that AI models will be kept secret.

In any event, that horse has already bolted. Many large language models have already been open-sourced, including models from American companies, like Meta's Llama models, and Chinese ones, like Alibaba's Qwen models, as well as others like Falcon 40B, MPT 30B and dozens before them. In many ways, seeing the AI race as somehow analogous to a cold war rivalry between two great superpowers misunderstands its nature. Yes, there is an aggressive rivalry over the hardware and the supply chain, and some models are being developed in proverbial bunkers in both hemispheres. But much of the technological development is also happening in the open, with open-source models available to developers everywhere and research being published for all to see.

That's a good thing for the world – especially those parts of it that don't have access to the vast resources and talent necessary to compete with Silicon Valley and the Chinese tech giants. Across the global South, for example in large swathes of Africa, Asia and Latin America, open-source AI models provide the only meaningful opportunity for local people and economies to utilise the technology, and ultimately benefit socially and economically from it. Generative AI is only going to have the sort of transformative effect that people hope it will have in many parts of the world if people are using AI tools designed with the needs of their local societies in mind.

Take, for example, Jacaranda Health and its AI-enabled digital health service, PROMPTS, which is designed to help expectant mothers across sub-Saharan Africa. PROMPTS sends women text messages in Swahili tracked to their stages of pregnancy. It also has an AI-enabled helpdesk that uses natural language processing to triage and respond to their questions, and makes rapid referrals if a risk is identified. To do all this, Jacaranda developed an AI tool that understands Swahili, built on Meta's open-source Llama 2 model. If people in sub-Saharan Africa are to realise the efficiencies, productivity gains and opportunities created by generative AI tools in the years ahead, it won't be by building and training their own expensive and energy-intensive AI models from scratch – it will be by adapting advanced models that have been made openly available.

A common but mistaken assumption is that releasing source code or model weights – the numerical values that guide the internal decision-making process of an AI model – always makes systems more vulnerable. On the contrary, it means thousands of developers and researchers can identify and solve problems that teams holed up inside company silos would take much longer to do. By seeing how these tools are used by others, in-house teams can learn from them and fix vulnerabilities. Researchers tested Meta's earlier large language model, BlenderBot 2, and found ways it could be tricked into remembering misinformation. As a result, BlenderBot 3 was more resistant to it.

Understandably, nation states will seek to monopolise some of the most advanced models for national security and military uses – whether on the physical battlefield, conducting cyber-attacks or defending themselves against attacks from their enemies. So it's prudent to assume that not every nation state will be happy to give open access to its advanced models for the greater good of all mankind. At some level, there will be the perceived need to hold some specialised advanced systems back in order to have a competitive and strategic edge over others. A similar incentive exists in the corporate world. Even a company as much in favour of open sourcing as Meta trains its models to create some systems and products that it doesn't release openly, because the company's intellectual property is its competitive advantage over its rivals.

Ultimately, however, openness is better for everybody than

closed systems. It leads to better products, faster innovation and a flourishing, competitive market. It allows for collaboration, scrutiny and iteration in a way that is especially suited to nascent technologies. It gives businesses, start-ups and researchers access to tools they could never build themselves, backed by computing power they couldn't otherwise access, which makes it the best way to ensure the opportunities these models create can be taken advantage of around the world. That said, while Meta's broad commitment to open source is partly philosophical, it is not all altruistic. As Meta's Chief Technology Officer, Andrew 'Boz' Bosworth, explains:

> When you open source something, a couple of things happen. One thing is you create a community around it, so it is self-reinforcing. Because the rest of the industry uses PyTorch [the Meta-built machine-learning library for building deep neural networks], a bunch of tools that we want are built not by us, but by other people that we get to use. And the things that we build are made better, more performant, more efficient by the community [. . .] If Llama continues to be what it is today – the center of gravity for a large community of builders – that's something we benefit from immensely.

In other words, people innovating based on open access to Meta's Llama models creates a feedback loop: developers, entrepreneurs and researchers all over the world build and improve their own products and systems which, in turn, helps the company improve its own products and systems – and therefore its revenues – over time as well. This cycle is supported by the company's advertising-based business model, which means it isn't reliant on charging people for access to its AI models in the first place.

That doesn't mean every model can or should be open-sourced – there's a role for both proprietary and open AI models – nor that models should be released willy-nilly without proper stress-testing. One way to think about the landscape of

AI models is with a graph like the one below, with four quadrants. The *y* axis indicates whether a model is for specialised-use cases or for general purpose. The *x* axis indicates whether a model is a closed system or one that is widely distributed via open source. In the bottom-left corner you have what will be the most closely guarded models – closed, specialised systems for sensitive areas of government activity, like national security and defence. Above it, general-purpose models like, OpenAI's GPT-4 and Google's Gemini, that are not released openly but are accessible for a fee. In the bottom-right corner are specialised models that are nonetheless openly distributed: for example, models that have been developed for specific medical purposes such as analysing medical images to detect cancer. And above that, in the top-right corner, are the openly distributed general-purpose models, like Meta's Llama and other open-source models.

There is a place in the world for models in all four quadrants. Each comes with its own benefits and risks. And while we can't eliminate those risks entirely, we can take steps to mitigate them. AI systems should be properly stress-tested. A common way of doing this is through a process called 'red teaming', in which teams – both internal teams and external cybersecurity experts – take on the role of adversaries to probe the integrity of systems by hunting for flaws and unintended consequences.

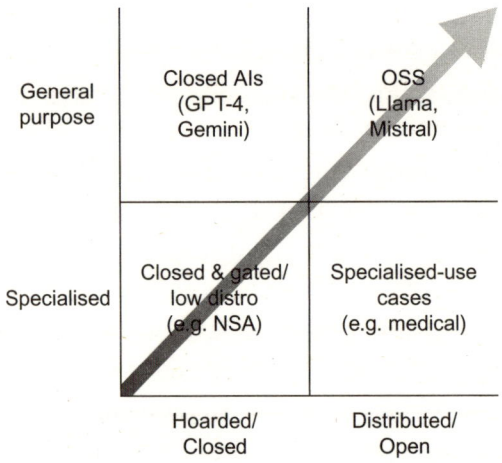

It's important to ensure there is a community of developers and cybersecurity experts with deep knowledge and experience who can get their hands on high-end open-source models – via the network of government-sponsored AI Safety Institutes in the US and elsewhere, or via the companies directly – in order to find weaknesses in them, build defences to them and improve their security even before they are released. For military-customised models, we need to adapt the existing conventions on weapons and warfare through the United Nations. The creation – and ratification by the US and others – of the Political Declaration on Responsible Military Use of Artificial Intelligence and Autonomy is an important first step. There is a clear risk to global security if we lack the diplomatic and moral leadership to set new standards and enforce them in the international community.

As future frontier models become ever more advanced – and potentially approach mythical 'artificial general intelligence' (AGI) status – there may be a stronger case for limiting the manner in which they are shared. But if done in a responsible way, with checks and balances in place to mitigate weaknesses, openness is surely preferable to keeping these hugely important technologies within the confines of a handful of already powerful companies.

All of this has brought us to a pivotal moment. Both geopolitics and domestic politics are reshaping the technological landscape. Diverging national internet laws and a wave of digital protectionism in democracies and autocracies alike are causing the global internet to fragment. America's support for the open internet it built is wavering. The US and China are locked in an escalating AI race that will have unpredictable implications for the development and spread of these powerful new technologies. While the open internet was developed in the era of globalisation, driven largely by American individuals, institutions and companies that saw the benefit in creating an open network based on the free flow of information through compatible infrastructure, it seems vanishingly unlikely that AI will be developed in such a collaborative manner.

Indeed, without global cooperation on the standards underpinning powerful new AI technologies, they could be fragmented from the start. The global economy could morph once again, hobbled by new digital borders and distorted by the huge disparities in both the nature and accessibility of AI-based technologies. This fragmentation would almost certainly have a big impact on how easily we connect with others around the world, significantly reducing our ability to collaborate and share knowledge across borders, in business, medical and scientific research, policing, academia, journalism and much else.

All is not yet lost. We can't restore the age of globalisation or the utopian idealism of the internet's early pioneers. But we can address many of the concerns and fears people have about the impact of new technologies and we can salvage some of the most important elements of the open internet. It won't be easy, and it won't happen without serious political will and concerted, coordinated effort. And it certainly won't happen without humility and openness from the big technology companies themselves. But it is possible. We need a new relationship between Big Tech and nation states in order to ensure workable guardrails are built around new technologies without killing the golden goose that has brought so much prosperity and opportunity to our societies. And we need the democratic world to come together to keep data flows open, to create standards and norms around AI that enshrine democratic values in these systems, and to make sure their benefits are spread around the world. For that to happen, we need to find a way to lower the temperature of the techlash, by bringing accountability to the Big Tech platforms and restoring trust in the idea that technology serves society. And we need the big democratic planets – particularly the US – to recognise that their enlightened self-interest lies in working together to share AI and keep data flows open.

PART FOUR

A New Balance of Power

CHAPTER 10

Let in the Sunlight

These days, politicians aren't held in universal high esteem, but I know from brutal personal experience that their life isn't easy. Everyone in politics is trying to be someone and do something. You want to win, you want to change law and public opinion in line with your world view and the causes you care about, you want to be admired and respected, and you want to leave a legacy. You can't stand up in front of thousands or even millions of people and say 'Vote for me' without a fair amount of ego and self-importance. But you also can't do it without some sense of mission or philosophy, and some idea of how to make the world better. Politicians get a bad rap because, in many cases, the self-importance is much more obvious to people than the sense of mission. But in very few cases are the motives entirely selfish, and in many more than you would expect, commitment to public service really is the driving force.

It's a life full of sacrifices and compromises, lived facing the constant glare of the spotlight and a steady stream of criticism. You are in a constant popularity contest – in elections, in the media, in opinion polls, within your own party and your own constituency. You lead a double – and sometimes triple – life, living in multiple places and trying to balance family with the public stage and the responsibilities of serving a constituency, a party and a country. You are stressed, overworked and under-slept, which can take a big toll on your family, relationships, and mental and physical health.

There will be areas of public policy that you are knowledge-able about and motivated by, but you are required to have a view on absolutely everything – from macro-economics, war and the climate crisis to the state of the potholes in the town centre and the frequency of local rubbish collections. If you hold office in government, you have the apparatus of the state at your disposal – and all the perks that come with it – but get-ting it to do what you want is a constant challenge, because bureaucracies tend towards groupthink, turf wars and inertia.

It is all too easy, with these competing pressures and the roadblocks in your way, to opt for the political rewards that come from loudly pinning the blame for a problem on someone else and demanding that they fix it, rather than doing the hard, complex and usually unrewarding work of actually trying to fix it yourself. So you ramp up the moralising rhetoric, demand some form of crowd-pleasing but non-specific action to bring to heel whoever the villains of the week are, enjoy the glow of the newspaper headlines, then hope someone else will engage with the details that might lead to a solution. Right now, politicians everywhere feel the pressure to 'do something' about Big Tech. And something must indeed be done. But what?

The debate around technology has become distorted, dishon-est and anti-science – and Big Tech is as much to blame for that as hyperbolic critics and reactionary politicians are. Platforms are responsible for connecting the world. But they also con-centrate large amounts of wealth and power in the hands of a small number of global companies. And when those companies are perceived to be innovating in their own interests and not in society's, this is toxic. As I said in this book's introduction, tech-nology has to serve society, not the other way around.

This accrual of economic and social power, coupled with years of denial and resistance to scrutiny, has led many people to believe the worst about both the technology and the technologists, has fostered a cottage industry of evidence-free moralisers in academia and journalism, and has encouraged troubled legacy industries to apply pressure to a political class that is just as confused by the dramatic pace of change as everyone else. And the consequences, as set out in the previous chapters, could be huge. Can we reset the debate before it's too late?

The techlash is fundamentally about power. People do not believe that technology companies use their power in the best interests of society. People are understandably suspicious of the scale of the platforms and the reach they have into our lives. They worry that many of the problems plaguing society may result from new technologies that they think are not being developed in their best interests. Parents worry especially, given the amount of time kids spend on devices and apps made by these companies. Politicians feel control ebbing away from them as decisions affecting their constituents are taken on the other side of the world.

This mistrust is not going to go away. If anything, the era of generative AI provides all the ingredients to make it significantly worse. There is likely to be an ever more extreme bifurcation between power being centralised in the few companies with the means to build and operate huge foundational AI models – effectively running a wholly new and borderless layer of energy-guzzling internet infrastructure – and the intimate, personalised experience that AI assistants will offer people in their daily lives. In some ways, the Microsofts and Amazons of the world are now embarking on the path that has made Meta so controversial. Social networks provide highly personalised experiences based around speech. This is why social media has been such a lightning rod for angst over digital technologies. The highly personalised relationships we will have with generative AI bots could take this to another level.

As I set out at the beginning of the book, politicians and other public critics peddle three fallacies about social media and how they operate, which lead to inefficient and at times harmful regulatory proposals. The first is techno-determinism: ascribing to technology an extraordinary capacity to make people think and do things without them having any choice or agency. In general, questions such as 'Does social media polarise?' or, more broadly, 'Does social media cause such-and-such a social phenomenon?' are simplistic and reductive. Presuming that social media is the main or only cause of any social phenomenon hides from view other crucial factors that shape individual opinion – such as family relations, socioeconomic status, race, gender, party affiliation, personal experiences and characteristics, education and so on.

The second fallacy is that correlation equals causation – in other words, the belief that, when a negative trend in society correlates with the rise of social media, the latter necessarily causes the former. But just because the growth of social media overlapped with the growth of polarisation in the US, that doesn't mean that one caused the other. Looking at the counterfactual hypothesis helps clarify things: if social media is this bad, then why haven't we seen similar trends in polarisation in all other countries with high social media use?

The third fallacy is that perfect content policies, not to mention perfect enforcement of them, are possible. The key trade-off for any content platform is between freedom of expression and safety. Both concepts are shaped by subjective opinion, culture and local history – in some places they are also shaped by law and constitutions, while in others they are not. Turning up the dial on freedom of expression necessarily makes it more likely that someone will be offended or, worse, feel unsafe. And freedom of expression and safety are both concepts which need to be applied at great speed to vast amounts of content on platforms used by billions of human beings around the world. There is no perfect content-moderation recipe. The assertion that it should be 'obvious' to the platforms where to draw the line is another great fallacy of the backlash against social media, and we could easily see this magnified in the era of generative AI. But we must remember that these are all just fallacies.

So, allowing for all of that, what can and should be done to *actually* address these complex problems?

A *healthy relationship between tech and nation states*

When powerful companies have erupted in size in a relatively short space of time, some politicians will react by calling for them to be broken up. The idea is that if companies are broken up into smaller pieces, consumers could benefit from greater choice.

The problem with that approach is that it completely ignores the benefits users gain from large network effects. People will tend to congregate around platforms where they can find most of what they're looking for – friends, content or both in the case of social media platforms like Facebook, Instagram, Twitter/X, TikTok and Snapchat; information in the case of AI chatbots like ChatGPT or search engines like Google and Microsoft's Bing (which are increasingly incorporating AI into their search systems); consumer products in the case of Amazon, eBay and Etsy; low-cost and easily available taxis in the case of Uber and Lyft; cheap properties for short stays in the case of Airbnb and Vrbo; movies or television shows in the case of streamers like Netflix, Disney+ and Amazon Prime; and many, many more.

The traditional logic of antitrust law doesn't work. These companies act as a marketplace where users and suppliers interact. For the many platforms that people flock to online, far from driving up prices and lowering standards, the reverse is true. The bigger the marketplace, the lower the prices and the better the user experience. Innovative new entrants aren't excluded – just look at the huge scale TikTok achieved over a short period of time, or the rush of users who embraced Chat-GPT when it launched. And scale has other advantages, such as the vast resources necessary for security and content moderation. Smaller platforms have fewer resources to invest in keeping people safe.

Breaking individual platforms up won't make the dilemmas and concerns that exist around content and safety easier to resolve. It will just lead to an endless game of whack-a-mole as regulators attempt to hammer platforms every time they achieve scale. And temporarily smaller platforms won't make the experience better for users or the market better for the millions of other companies that benefit from the platform economy.

But accepting that old-fashioned antitrust actions won't address the fundamental dynamics of the internet economy doesn't mean the status quo is flawless either. The European approach to regulation places onerous compliance obligations on big companies, some of which, as I described earlier, deter companies from making products or features available to Europeans. But its intention – in theory, if too rarely in practice – is not to cut big aggregating platforms into smaller chunks, but to legally enforce greater transparency, accountability and sovereignty for data citizens in how they interact with these big platforms. This is philosophically distinct from the US agency-led attack on size, but it goes with the grain of how data-rich platforms work (in theory, at least).

One way to make the theory work better in practice is to base any new regulation on an evidence-based analysis of the problem, rather than on a politicised desire to cut Big Tech down to size regardless of the practical realities. It is here that EU legislation has notably failed, time and again, imposing requirements with little merit for consumers or the economy, driven unduly by an evidence-free desire to impose 'sovereignty' over (predominantly US) tech platforms. But in order to achieve this evidence-based analysis, we need to arm politicians – and the armies of civil servants, advisers and think tanks that support them – with the data and research that can inform the policies designed to hold tech companies to account. Only if we equip them with the information they need can we expect them to take responsibility and engage seriously with workable regulation.

The Wild West days of the internet are long gone. The days of moving fast and breaking stuff are well behind us. The last two decades or more have created a powerful new class of tech billionaires; through their companies and products, they now exercise huge influence over the public sphere. Governments come and go – some more pro-tech, some more anti – but the trend is clear: one way or another, politics will continue to assert itself over technology, responding to a widespread societal demand that the power and influence of private technology companies be held to account. As Tony Blair told Mark Zuckerberg when I first introduced them: 'You basically have a choice – get broken up or get regulated.'

Benedict Evans describes the three ways tech companies – and, indeed, any big private sector companies – say no when faced with regulations. The first is to object to regulation simply because they don't like it. The second is to object on the grounds that the proposed law is a bad law, because it might lead to consequences the company understands but that policymakers haven't quite grasped. And the third is to object because the requirements imposed are technically impossible to implement.

The problem for policymakers is that it is often hard to know which 'no' they are hearing. Is the tech company saying their proposals are flawed or impossible to implement because they really believe it, or because, deep down, they just don't like them and want them to go away? In the absence of any trust or goodwill in the relationship, why would the policymaker accept that the company's arguments are made in good faith?

Tech companies have to resist the urge to say no the first way. Obstinacy won't make regulation they don't like go away; it will only increase the appetite for it. They can reduce (but never eliminate) the likelihood of being faced with regulation that is fundamentally flawed or that would lead to bad consequences. But if they want better-informed regulation, they will need to give out better information to enable it.

Policymakers in turn have three stock responses of their own that they often revert to when faced with technological change. The first is to simply try to stop the change from happening, often by siding with vested interests whose power is threatened by technological disruption. They might, for example, take the side of the taxi lobby against Uber, or side with local hotel groups against Airbnb.

The second is to shift responsibility. New technologies often create difficult trade-offs and dilemmas around timeless political tensions like privacy and security. To go back to Benedict Evans's earlier example, a case in point is encrypted messaging services like WhatsApp. No amount of 'nerding harder' can make private messages readable by government and law enforcement while not being readable by foreign spies and malicious hackers. You can't invent your way out of conflicts between values. By passing the buck to tech companies, politicians are washing their hands of a responsibility that is theirs. We elect political leaders to resolve political questions about the rights of citizens and the collective trade-offs for society.

The third response is to get the new to prop up the old. As a 2018 report by the Tony Blair Institute for Global Change notes:

> This response is most common when the tech companies are perceived to be highly profitable, which makes it easy to mount a populist argument for redistribution. And so UK Labour Party leader Jeremy Corbyn calls for a tax on Facebook to pay for local journalism, and on Netflix to pay for the British Broadcasting Corporation (BBC). UK Chancellor of the Exchequer Philip Hammond wants a tax on Amazon to prop up high-street shops. US President Donald Trump also wants to hit Amazon by making it pay higher fees to the US Postal Service, and looks to have Google in his sights next.

What does a healthy relationship between tech and nation states look like? Let's start with what each side is good at. Private tech companies are good at innovation and value creation. Nation states are good at setting the rules on everything from privacy to youth protection, maintaining open markets with fair competition, protecting critical infrastructure, ensuring that disadvantaged groups can access technology, distributing equitably the wealth created by tech (including AI productivity gains on the horizon), and directing investment for social ends. What we need is a system in which both nation states and private companies play to their strengths.

It is in the self-interest of tech companies to be open about their processes and systems and take responsibility for what comes of it. It is in the interest of societies that policymakers react to that openness thoughtfully and constructively. Only with much greater transparency and accountability from Silicon Valley can the political world have the debate it really needs if it is to respond with proportionate and evidence-based policymaking. How much risk are we prepared to accept for progress?

That starts with identifying as best we can what the real risks are – in terms of both the impact that existing technologies have and the potential risks that new technologies could pose – especially when it comes to the health and welfare of children and other vulnerable people. It means examining the sorts of guardrails that are possible and desirable, and who should be responsible for imposing and policing them. And it means weighing those risks against the benefits that will be lost if unduly strict boundaries are imposed. So Silicon Valley is going to have to throw open the curtains and let the sunlight in.

Letting the sunlight in

A radical level of transparency from big internet platforms would be good for everyone. But we're only going to get that if all companies are held to the same standards, and if this transparency and what it reveals is equal in some way across the whole industry. One company revealing its secrets while a rival doesn't puts the more transparent company at a disadvantage. And if it is left to individual companies to decide how transparent to be, their decisions are vulnerable to the whims of those in charge, to paranoia about openness increasing vulnerability to attack or criticism, and to cost-cutting. Which means that if we want a meaningful standard of openness across the whole sector, it can't simply be left to the goodwill and foresight of the tech companies themselves.

The only way to safeguard transparency is to write it into law. As some of the more sensible new regulations recently introduced around the world demonstrate, transparency and accountability can be underpinned by legislation, with the threat of hefty fines and other punitive actions if platforms fail to comply. And while hefty fines provide regulators with a big stick, there's a carrot too. Equal standards and comparable disclosures across the industry mean market forces can push companies to improve their products and practices.

As Stanford academic Nate Persily told a Senate subcommittee in 2022 when questioned about a bipartisan bill that he helped to develop – the Platform Accountability and Transparency Act (PATA) – to force tech companies to offer up much more data for public research, transparency forces companies to change: 'I get criticism that it's weak legislation because it's not breaking up the companies or it's not going right after content moderation. But once the platforms know that they're being watched, they will change their behavior.'

I've seen at first hand the galvanising effect of regulation on private companies, with huge internal operations created in order to ensure compliance with new legal requirements. In 2019, the Federal Trade Commission fined Meta – then Facebook – $5 billion for privacy violations and imposed a series of transparency measures on the company, akin to the sort of mandatory reporting big financial institutions are subject to. Among other things, the company was required to create new roles and processes to safeguard people's privacy, disclose extensive information about privacy violations, publish independently audited reports around its approach to privacy issues, and submit itself to regular independent assessments to ensure it was complying with the requirements of the order. Meeting these obligations required a huge internal effort to rewire the company's processes and decision-making. But the company did it. It may not have had any choice in the matter, but more than half a decade on it is clear there has been a change in the company's culture. How it protects people's privacy and their personal data, and the controls it offers people to ensure they can manage their privacy settings, are now baked into the company's product-development processes from the word go. The glaring problem, of course, is that these stringent requirements do not appear to have been imposed on other major platforms that are handling people's data at scale in the same way. The approach is both partial and incomplete. Sensible regulation is only as good as the evenness of its application.

With that in mind, here are some steps that all major tech companies can take to become radically transparent, and that lawmakers could consider writing into the statute books.

1. Opening up the books

During my time at the company, Meta did far more than most to lift the veil on its own internal processes. The company publishes its policies and distribution guidelines, quarterly reports disclosing its success rate at taking down malicious online campaigns and other security threats it identifies, and quarterly reports disclosing the prevalence of hate speech and other forms of content that violate its policies, which are externally audited by EY. Meta's teams log government takedown requests publicly in the LUMEN database, an independent research project from Harvard's Berkman Klein Center for Internet and Society which studies takedown notices and other legal removal requests and demands concerning online content. Internal Meta researchers collaborate with academic researchers through bespoke projects like Social Science One and the US 2020 project that examined the impact of Facebook use on political attitudes, and the company provides tools for researchers to access data in a way that protects individual privacy through the Meta Content Library, which allows researchers to search publicly available content on Facebook and Instagram, and the Ad Library, which publicly archives every paid political ad posted on Facebook and Instagram for a period of seven years. And Meta publishes what it calls 'system cards', which give insight into how its systems work – including the AI systems that rank and recommend content – in a way that is accessible for those who don't have deep technical knowledge.

It is easy to imagine the sort of externally audited reporting about privacy, security threats and content removal that Meta does (both voluntarily and in response to the FTC order) becoming a legal requirement across the industry, potentially alongside other types of data disclosures deemed by regulators to be in the public interest. Legislation could describe the sort of content that should be archived and made available to governments and researchers, perhaps including content that platforms remove so that allegations of 'censorship' can be properly investigated. States could insist on routine and ongoing collaboration with researchers to test hypotheses and get causal findings on issues of social concern – youth wellbeing, democracy, filter bubbles and so on – with companies required by law to respond to the findings detailing how they are building safeguards into their products.

2. Devolving decision-making power

If one of the biggest concerns about the power of big internet platforms – and social media in particular – is that they exercise unaccountable power over public discourse, one simple remedy is to introduce checks and balances on the controversial decisions they make about what content is and isn't allowed on their services. I long argued inside Meta that, as an innovation company, it ought to be applying that mindset not just to its products and emerging technologies, but to new forms of governance too.

One model for that exists, albeit on a relatively small scale. In 2020, I oversaw the establishment of what was then Facebook's independent Oversight Board, an idea first proposed by Mark Zuckerberg in 2018, which I described earlier as a sort of Supreme Court for the most difficult content decisions. The experts on the Oversight Board hear cases about controversial content that are referred to it by individual Facebook or Instagram users, or by the company itself, and make binding decisions about whether the content should be taken down or remain up (or be reinstated if a previous decision to remove it is being challenged). It can also make non-binding recommendations about changes it believes the company should make in its policies and enforcement practices.

It was always an ambition for both Meta and the Oversight Board itself that its role could expand beyond just one company to become an independent adjudicator for social media companies more broadly. The board's model has certainly caught the eye of regulators and industry rivals. The EU's Digital Services Act includes a requirement for independent appeals boards for social media companies, leading to the establishment of the Appeals Centre Europe, an out-of-court settlement body where users can appeal against policy-violation decisions from major platforms. This is a separate entity, but it has been set up with the Oversight Board's support. And the board has inspired other tech firms, like Spotify, which has created its own Safety Advisory Board to review claims about harmful content on its platform.

3. Bringing users into the process

Another radical approach to platform governance could be what are known as community forums. Meta has experimented with these forums to bring people together to discuss tough issues, consider trade-offs, and share recommendations for improving people's experiences across our apps. The model is based on deliberative polling, an approach that has been used by governments around the world, in which representative groups have the opportunity to learn about complex issues before sharing their perspectives. This differs substantially from more typical user-experience surveys, in which people are polled for their opinions but aren't necessarily familiar with the subject matter. Participants in community forums have access to extensive educational materials, deliberate multiple times in small groups, and are given opportunities to ask questions of experts about the concepts discussed. This helps them engage more deeply with complex issues and leads to more considered and nuanced results. And it helps those in charge – whether they are policymakers or private companies – to take on board a wide range of diverse perspectives from users or citizens as they design policies or build systems and products. It is a model which derives from the ancient Athenian idea of sortition, in which lotteries were used to ensure public-office holders were representative of the citizenry (or, at least, the section of it made up of men over the age of thirty). Its closer cousin is the citizens' assembly, a representative group selected at random from the broader population to learn about, deliberate upon and then make recommendations on a given issue or policy area. These have been used by devolved governments and local authorities in the UK in recent years, for example as part of a House of Commons committee inquiry in 2019 into the long-term funding of adult social care in England.

Generative AI is well suited to this method of decision-making. AI models are guided by the data they have access to, as well as by the structures and inputs in their design. Meta and other AI companies input 'values' which guide AI models and can help protect against bias and unintended consequences by giving the model a way to evaluate its own outputs. Community forums can help ensure these values reflect different viewpoints from throughout society, not just those of software engineers in Silicon Valley.

Deliberative democracy mechanisms like these have been used by governments and organisations around the world for decades to answer difficult questions: from amending the constitution in Ireland to addressing environmental disasters and population pressures in Uganda, and changing the election system in parts of Canada. In recent years, Meta has experimented with community forums as a means of getting input from users about policies in a range of areas, working closely with Stanford Deliberative Democracy Lab and the Behavioural Insights Team (BIT). In 2022, a series of pilots in Nigeria, India, Brazil, the US and France tested whether community forums could be helpful for advising on what Meta should do about misleading climate content, giving the company a much better understanding of how people felt about this tricky content-moderation issue. Later that year, a bigger forum was held of nearly 6,000 participants across thirty-two countries to hear people's ideas about personal conduct in virtual reality spaces, in order to inform Meta's policies as it develops and expands the metaverse platform Horizon Worlds. Participants were able to read educational materials and question security, privacy and social media experts before taking part. Then, in 2024, a pilot about generative AI, involving 1,500 people from Brazil, Germany, Spain and the US, found that the participants were generally positive about AI, that they felt chatbots should have the ability to remember past conversations in order to improve their performance as long as the people using them are informed, and that they were in favour of AI chatbots being human-like – again, as long as users know that they aren't dealing with a person.

4. Giving users meaningful control

How can people feel that technology really serves them, and not just the companies providing the technology in the first place? Community forums give people the chance to participate in policy choices, but they don't give them direct control over their experience online. If transparency is essential for legitimacy, control is essential for accountability. People need to feel they are the ones in charge of their relationship with technology, that their rights are being respected, their expectations being met, and their choices being responded to.

This is especially true of people's relationship with data. Increasingly data-hungry AI models require an exponentially increasing supply of data for training. And it is likely that a wide variety of types of data – from text to biometric – will be needed for higher-quality models in the future. When you consider the sort of information you might need to disclose to your AI assistant for it to provide hyper-personalised services – from life planning to therapy sessions – it's clear that we need to be thinking urgently about how to give people meaningful controls over the way their data is used and stored.

A crucial element of this is giving people a range of options for how their data is used in different contexts. The scholar Helen Nissenbaum coined the phrase 'privacy in context' to describe the way norms and expectations around the sharing of information differ depending on social context. In her book of the same name, she gives the example of your relationship with your doctor – a situation in which you are prepared to share information you probably wouldn't choose to share with your friends. AI assistants will have a wide range of different uses, so there will be different expectations, and therefore different rules and norms established, in different cases.

To exercise control over the use of personal data for these tools, people will first need to understand what sort of data is required and where it will be stored (for example, whether on their personal device or a company server), what options they have (for example, whether there are data-light or data-intensive versions of the experience, and whether data can be deleted once the experience is over or whether it is necessary to store it). In practice, this will probably mean prompts within the apps that set out the context and choices available to users, as well as reminders and other visual signals embedded throughout the experience. Giving parents the ability to set these controls for apps used by children or teens will obviously be essential too.

This is something Meta has tried to do over time in its social media apps. People can choose to revert to a chronological feed on Facebook or Instagram if they want to opt out of the algorithmically ranked feed. The 'Why am I seeing this?' feature gives people information about the signals that were used in Meta's content-ranking systems to explain why a post has appeared in their feed. There are centralised places on Facebook and Instagram where people can customise controls that influence the content they see on each app. The 'Show more, show less' feature on Facebook and the 'Not interested' feature train the algorithm not to show people types of content they're not interested in.

Meta's teams try to anticipate people's reasonable privacy expectations and bake them into products at the development stage – an approach it calls 'privacy by design'. So, for example, if people behave aggressively or inappropriately in metaverse spaces, people should reasonably expect to be able to report this bad behaviour without a record of their conversations needing to be stored indefinitely on a company's server. For Meta Quest users in Horizon Worlds, a rolling buffer is available so that most audio data can be kept for just a short period to ensure it is available for users to report abusive or harmful conduct. If they don't submit a report, the data is deleted. Similarly, people would reasonably expect that if they interact with an AI assistant inside a messaging app, that assistant should only be able to see things people have chosen to share with them and not have sweeping access to every otherwise private message they've sent on the app. So for people interacting with Meta AI within Messenger, Instagram and WhatsApp, the AI can only see what the users choose to send it: direct one-to-one chats with the AI, messages that tag @MetaAI, and user feedback are shared with Meta. These generative AI models are not trained on people's private chats with friends and family. Users can also delete their prompts to AIs using the same deletion controls they already use for other messages.

Another way for users to exercise control over their data is through what's called data portability – the ability for people to transfer their data from one app or product to another. If you use an AI assistant built by one company but want to switch to another, you may well want to bring the personal data the former has learned from to the latter. By definition, this can't be done by companies in isolation. Someone who wants to take their information from a Meta chatbot to an OpenAI one is going to need both companies to have worked out how that data is shared securely. This is an area where the law could enforce standards across industry. And it is probably better that policy-makers lead here, because you can immediately see a tension: if the data one company holds on one person is intrinsically linked with that of others – as is inherently the case with social networks – but privacy laws like GDPR enshrine the principle of personal ownership of data, then how can we ensure people can move their shared data from one service to another without jeopardising the privacy of others? In other words, how do we strike the right balance between rules governing the movement of data and the rules governing the privacy of data? That's a political judgement rather than a technological one.

Laws could also require measures to help users understand changes to products that affect their experience. For example, when a significant change is made in the algorithms that dictate the content you see, or when a new product is launched, plat-forms could help users understand those interactively through a conversation with an AI chatbot that walks them through the changes, rather than simply posting the announcement on their company websites.

There is also a host of innovative ideas springing up to give people more control over how their data is used and stored. 'Data trusts' are markets of independent third parties that store and manage user data outside social media platforms. These third parties interact with social media platforms – and in this case with social media platforms' AIs – in ways that protect access to personal data. So, to borrow the *MIT Technology Review*'s example, groups of Facebook users could create a data trust to determine under what conditions they would allow the app to collect and use their data. The trustees could, for example, set rules about the types of targeting that Facebook could employ to show ads to users in the trust. If needed, the trust could retract the company's access to its members' data.

Similarly, the political scientist Francis Fukuyama has offered the idea of 'middleware', which is 'software that rides on top of an existing internet or social media platform such as Google, Facebook or Twitter and can modify the presentation of underlying data'. This would create a bespoke interface between the user and the app that allows the user to tailor the content that appears, in effect allowing people to apply their own personal content rules. Middleware would give users more control over what data is shared, and how, with these platforms, effectively enabling them to personalise their own experience of these apps. That said, middleware raises a whole host of issues around data privacy and would require high user participation to work well. As presently framed, it is not an idea which is likely to work in practice. And if platforms build much more user-friendly controls into their apps in the first place, then the case for third-party programmes like these would diminish further.

There are also interesting forms of decentralised networks emerging that could provide alternatives to people handing over large quantities of personal data to individual companies. Blockchain is a decentralised database built on the principles of anonymity and user control. While every transaction in the blockchain is publicly visible, it is cryptographically protected so no one can access that information other than the owner of that transaction. Blockchain could become the platform that chatbots engage with in order to elicit information about their users – and do so in privacy-protected ways.

Another decentralised network concept that is taking off is

the Fediverse, which, if you can get past the name, is a loosely connected 'federation' of networks that communicate with each other, but which have their own rules as to how data is stored and processed. Meta's social app Threads is connected to the Fediverse, which means Threads users have an open connection to other social apps like Mastodon, despite both being separate networks. Mastodon users can see content on Threads and vice versa, but how their data is used will be governed by the platform they signed up for. So in the Fediverse you can use the social app of your choice, but get content from any of the others that are connected to it. And with no central authority overseeing the Fediverse, it is more resilient to censorship and corporate influence. It is at least theoretically possible that this approach could be extended to chatbots: those who use a Meta chatbot could thus have their chatbots interact with OpenAI chatbots without having to accept the terms of service and data policies of OpenAI. And vice versa.

The uncomfortable truth for Big Tech

Right now, some of the largest platforms are betting big on AI and spending tens of billions of dollars on the hardware, talent, research capabilities and product development necessary to compete in a race for a lucrative share of what they believe will be a truly transformative new technological era. But many policy-makers all over the world share the view that AI technologies may simply be too powerful to be left in the hands of a small number of private sector companies, particularly when they have significant concerns about how those companies operate and significant political incentives to treat them with scepticism. Regardless of the pros and cons of digital nationalism, nation states are steadily expanding sovereign control over the laws of the internet, over their citizens' data, and increasingly over domestic AI capabilities. Tech companies will have to face up to the risk that the AI infrastructure they operate – the data centres, high-power computer chips and even the foundational AI models themselves – could come to be seen as a vital public utility that should be brought, one way or another, via regulation or ownership, under public control. That is what has happened to privately built infrastructure in the past, and it could happen again.

Governments are hardly short of political, ideological or geostrategic incentives to do so. Sovereign AI capability is not only an economic imperative for governments, but it has big implications for their national defence and the administration of public services. The US and China are locked in an increasingly aggressive trade war, in which access to AI chips is a key component. Modi's India has long been developing its own nationalist industrial policy. We've seen growing political support for 'national AI champions', like Mistral in France or Aleph Alpha in Germany, or the establishment of a national AI investment agenda in the UK. And given the huge energy demands that AI data centres require, states are beginning to look at them through the lens of their climate change objectives too.

As I say, there is plenty of precedent for the state taking control. Much of the infrastructure of modern societies is either state-owned or licensed to private sector companies by the state. At various points in the last century, the US has nationalised the airwaves and licensed access to the public spectrum to private broadcasters; taken over its largest train company to create what became Amtrak; and bailed out banks and General Motors. In 2018, the Trump White House looked into whether it should nationalise parts of the 5G network. In the UK, the National Health Service is a cornerstone of British society, and at various points in the twentieth century the state has nationalised water, energy, the railways, aerospace, telecoms, coal, oil and gas. Prime Minister Keir Starmer's Labour Party was elected on a platform of creating a state-owned energy company called Great British Energy. Big Tech may have sunk billions upon billions of dollars into creating, developing and building the foundational layer of this new technology, but that doesn't mean it can't be taken out of their hands.

That would be a bad outcome for all manner of reasons. For starters, it's a surefire path to fragmented technological development and the slowing of innovation. As the Carnegie Endowment for International Peace notes, one of the 'potential pitfalls of pursuing a data sovereignty strategy' is an inability to maintain 'access to high-quality global data' that is 'necessary for developing effective AI-training algorithms'. Moreover, and more importantly, it opens the door to state control, surveillance and censorship. And it crystallises the AI race as adversarial, not only between the US and China, but also between every nation state that seeks to establish a significant domestic AI sector. It's a recipe for escalation rather than cooperation, with nations pouring mounting financial and political resources into the abyss as they try desperately not to fall behind their rivals. Ultimately, it would lead to a technologically divided world of AI haves and have-nots, in which countries that can devote vast resources to infrastructure, research and innovation pull away from those that can't.

In the next chapter, I'll propose a way for governments – and particularly the world's biggest democracies – to avoid such a dismal outcome by channelling this political energy into the international sphere. But Silicon Valley can't afford to be its own worst enemy here. Something needs to give. If Big Tech wants to avoid having its golden goose seized by the state, it can't just sit back and hope that policymakers see sense.

Tech companies need to appreciate where their enlightened self-interest lies in all this. They may never be truly trusted again as they have been in the past – by the governments who regulate them or the people who use their services – but they can be far more open and accountable. If they are, they can lower the temperature of the techlash and remove themselves from the firing line; they can be a valuable partner for governments; they can play a mature and responsible role in society; and, crucially, they can retain society's permission to innovate.

However painful it may be for them in practice to significantly increase their transparency and accountability, especially when it's forced upon them by regulators, what is fundamentally at stake is their licence to operate. If Big Tech companies stay closed-off and refuse to provide data for the purposes of accountability, experts and governments are less likely to understand their systems, less likely to understand their impact in the world, and less likely to trust that they are able to regulate themselves.

As Nate Persily remarked when giving evidence to the US Senate:

> We cannot live in a world where Facebook and Google know nearly everything about us, and we know next to nothing about them. These platforms have lost their right to secrecy, and it is well past the time that someone other than the firms' own data scientists be granted access to the data that reveal the impact of these platforms on the information ecosystem.

Letting the sunlight in means more than the Big Tech companies just being open about how their systems work and cooperating with governments and civil society. It means actively supporting public-interest research, even if that might lead to difficult questions and dilemmas for them. Transparency isn't painless. Nor should it be. True accountability means there will inevitably be times when the data companies reveal things that are uncomfortable and need to be fixed. In practice it would also mean giving their critics sticks to beat them with. And all the while, they can expect to get not a jot of credit for what they are doing. It's no wonder they are reluctant. But the self-interested trade-off for tech companies is clear: the pain of scrutiny and accountability is much better for them than being broken up; much better for them than being subjected to punitive and ill-conceived regulations; and much better for them than having their AI infrastructure nationalised.

Allowing third parties access to data from which they can draw their own conclusions is a vital part of ensuring that private companies remain in sync with the values of the wider world. Of course, tech companies will reserve the right to disagree with the way that people interpret their data. But when this happens, transparency allows them to defend themselves on the basis of shared facts, and it allows external researchers to critique and correct poor-quality third-party analysis.

None of this guarantees good policymaking. Some bad laws will be written, no matter what. But there's a reasonable chance that this kind of radical transparency will increase the likelihood of good, evidence-based laws that are practical to implement. Good laws that understand how technology works and minimise the likelihood of harm and unexpected consequences are in everyone's interest.

And of course there will always be politics. Governments will continue to assert national sovereignty, as they should. Borderless tech platforms present an obvious challenge to governments, which operate within borders. And there will continue to be real issues of harm – as there are with all technologies – that require scrutiny. But the relationship could be so much healthier for all concerned.

The uncomfortable truth for politicians

There is no magic wand which would make all online technology 'safe'. That's because, despite all the hype, it consists of tools. What matters is what people do with those tools. We have seatbelts and airbags in cars, driving tests and speed limits, and punitive penalties for drinking and driving. But some people still drive dangerously and, sadly, people still die in car crashes. Some people will choose to do bad things, some will be careless, and some will make stupid mistakes. We can – and must – legislate to reduce the damage that can be done, to create norms around good behaviour, and to stop people of good faith from doing things they shouldn't. But nothing, short of banning the internet itself, will ever make the online world totally safe.

What will still remain once new laws and regulations have

been written into the statute books are the age-old political dilemmas at the heart of so many of the debates about tech. Where should the line be drawn between the right to free expression and the right to share hate and lies, and to glorify violence? At what point should someone's personal privacy be violated in the name of the safety of their fellow citizens? Who bears responsibility for stopping bad people from doing bad things? And when they are found to have done bad things, should they be treated harshly or leniently?

Questions of this kind can never be completely resolved, because people will never completely agree. They are questions that sort people into liberals and conservatives, socialists and libertarians. Right now, tech companies are blamed for most of the ills of the internet, in large part because they are the ones making the decisions over questions like these, leaving them damned if they do and damned if they don't. They are both powerful and unaccountable, as well as human and fallible. But once politicians have set the rules, the responsibility passes to them. And they will soon find that you can have light-touch regulation and be damned when bad things happen, or overbearing regulation and be blamed for restricting people's liberties.

We can't wish away the problems inherent in giving everyone a means to express themselves, nor can we turn the clock back and uninvent technology. New rules for the internet have long been necessary. Unaccountable power needs to be made accountable. But once the laws are passed and we settle into our new status quo, many things will remain unresolved. The unrest of those left reeling in the wake of the financial crash will still be there, and they will still be angry. Young people facing the prospect of less secure careers and lives than their parents will still be there, and they'll still be angry too. We will still have millions of people struggling with their mental health. There will still be a climate crisis. Public services will still need to be paid for. Racism, hatred, misogyny and bigotry will still exist, as will war, disease and injustice. Big Tech has made numerous missteps, but it is blamed for a lot more than that. Some of the bad will be mitigated by new laws, but some of the good it does will be restricted too – although hopefully not too much. To be held accountable, it must become radically transparent, but the most important thing isn't opening up the books; it's what we do next with what we learn from that process. And at that point, the ball won't be in Big Tech's court any more – it will belong where it always should have been, in the realm of politics.

We will still face the twin threats of internet fragmentation and the AI arms race, and the prospect of a more divided technological world in the decades ahead. Neither is inevitable, but they are likely to happen if governments continue to act in isolation. So, while a much more open approach is in Silicon Valley's enlightened self-interest, it is also time for democracies – especially the world's three biggest techno-democracies: the US, the EU and India – to examine where their enlightened self-interest lies.

CHAPTER 11

A Deal to Save the Internet

As we have seen, among the world's most technologically advanced nations, the United States holds the strongest hand. But, as we have also seen, its pre-eminence is more fragile than it might appear. The global internet still works on its terms, broadly speaking, but other nations are now asserting their authority over the internet within their borders, chipping away at America's influence. US companies still operate the internet's biggest platforms, generating eye-watering sums of money for the US economy and attracting the best and brightest talent from around the world to America's shores. But other nations are offering significant incentives to their domestic tech industries and are trying to hobble America's tech giants: a number of big Chinese tech companies in particular are emerging to challenge the supremacy of Silicon Valley. And while American companies still have the lead in the AI race, aided by the US government's attempts to stifle China's access to vital AI infrastructure, China is navigating its way around America's trade restrictions and is catching up. Dozens of other nations are understandably intent on building up their domestic AI capabilities too, for fear of becoming second-class citizens in the new era of generative AI. So while America is leading right now, the balance of power could easily shift. If it is to consolidate its position as the world's supreme technological power – and the de facto underwriter of the internet's value system – America needs to reconsider where its enlightened self-interest lies.

The most dangerous thing for the US to do is carry on with business as usual. If it refuses to defend the open internet, the open internet will disappear, draining away America's cultural and economic hegemony. If it escalates its ever more protectionist approach to AI technologies, it will only incentivise rivals and allies alike to build up their domestic industries, with many likely to turn to America's biggest rival, China, for support. If as a result China's models, rather than American ones, become the default AI technologies used in Asia, Latin America, Africa and even Europe, then China's sphere of influence expands as America's recedes. And when that happens, the influence of American values and norms – so successfully exported during the internet's expansion – recedes too. Soon the internet, along with the wider landscape of digital technologies that pervade our lives, underpin our economies and are deeply embedded in our public services, will operate according to a different set of values, developed in a climate where censorship, surveillance and other types of top-down state control become more prevalent.

If America creates a vacuum, China will fill it. You only have to look at the rapid rise and pervasive influence of Chinese apps like TikTok to see how easy it is for America's rivals to spread technologies well beyond their traditional sphere of influence. To avoid this, American leaders need to recognise that their interest lies in rallying the world's techno-democracies – most crucially the EU and India – around a set of shared principles and arrangements that protect what remains of the open internet, and to share access to foundational AI technology. America has only a small and closing window of opportunity in which to act. If it does, the upside is enormous: it can in effect enshrine its liberal democratic values in this new layer of the global internet, and ensure that the cultural and economic influence America enjoyed in the first era of the internet continues into this new one. In doing so, it can maintain its technological supremacy and create a bulwark against the spread of the authoritarian internet model.

A shared sense of purpose based on common values like free expression, transparency and accountability could be the foundation for a US-led global consensus that governments, industry and civil society can organise around. By working together in a renewed spirit of shared endeavour, democracies could put guardrails in place to ensure AI is developed in a way that serves society and stops the fragmenting of the internet. But above all, encouraging free enterprise and protecting core democratic values like free expression must be inherent in the deal.

These are all fine sentiments, you may say. But what exactly would this shared endeavour consist of? I'm proposing a multi-layered deal that could be agreed between nations, composed of the following two key elements: open data flows to maintain the fundamentally borderless nature of the global internet, and arrangements that allow the basic ingredients of AI – the hardware, infrastructure, talent, research and advanced models – to be shared and utilised. To be effective, such a deal would need to have the world's biggest techno-democracies as signatories. That means, as a minimum, the United States, the European Union and India, but it should also include the United Kingdom, Brazil, Japan, Indonesia, South Korea, Canada, Australia and many others. In keeping with the network effects of the internet, the more nations are willing to sign up, the more effective it will be.

This may seem unrealistically ambitious in a world that is fast deglobalising. There are a thousand and one reasons for nations not to embrace such full-throated internationalism. That is why it is important that, while there must be practical and self-interested reasons for nations to sign up to the deal, it should also be modular, so that others can participate in elements that appeal to them in a way that is inclusive and not exclusive, perhaps with incentives for closer alignment in future. Even if the world is polarised, and even if China is increasingly adversarial, the West should always leave a door open. Collaboration and cooperation around issues of mutual self-interest offer the best way to de-escalate geopolitical tensions, even if it seems fanciful to imagine this sort of joint effort today.

The deal, part one: an open internet treaty

The first element of this deal addresses the threat to the open internet by keeping it connected so that data can continue to flow between nations. For that to happen, all the signatories to the deal would have to agree to a new multilateral model for data governance. What would that look like? First, we need to consider what each signatory wants.

The US wants American companies to remain the pre-eminent global platforms and therefore not to be blocked or limited from accessing local markets, and to maintain their lead over Chinese-owned competitors. And it is also in the US's interest to create a bulwark to protect free expression and other fundamental rights from the spread of the authoritarian internet model and, in particular, alternative technical architectures that could pose a threat to US security.

For America's allies, it's clear from the thrust of the data-localisation laws being considered around the world that governments are fundamentally looking for three things: sovereignty over what is and isn't acceptable in the public sphere of the internet within their borders; the ability to protect their national security; and a thriving domestic internet economy. It is not acceptable, in the eyes of many of America's allies, for US companies to impose content rules and – on the basis of policies written in line with American First and Fourth Amendment protections – refuse requests to take down content they regard as illegal in their own jurisdictions. For many nations, rules written by US companies based on US law and US standards constrain local law enforcement, limit their ability to communicate with their citizens, and restrain their ability to limit the spread of information within their borders that they consider contrary to their national interests – including matters of national security.

Whatever one's views on the merits of US constitutional principles, these are unavoidable political reactions that need to be respected in any new data-governance framework.

The solution is a new treaty that protects the free flow of data between signatory countries while upholding commonly accepted standards of democratic rights. That means the rights of ordinary citizens to express themselves freely in line with the protections enshrined in Article 19 of the UN's Universal Declaration of Human Rights; the right to know when a government is seeking to restrict your speech, and under what legal provision; and the right to challenge breaches of human rights in a national court or an international arbitration body. It also means the rights of countries to restrict speech in line with the provisions of Article 19: that the restriction is lawful; and that the grounds for restriction are specifically to protect national security or public order, public health or morals, or the rights or reputations of others. And it means the rights of governments to access data related to serious crimes, subject to reasonable justifications, workability and independent oversight. For private companies, it would enshrine their rights to challenge content takedown requests, to protect employees from threats, and to tell their users why they are restricting content and at whose request. Article 19's free speech protections may not be identical to the First Amendment's, which could prove an obstacle for a US administration determined to push for a comprehensive approach to free speech, but it remains a widely accepted standard and likely the best hope for maximising free expression globally.

The deal, part two: sharing AI

The second element of the deal is AI. Again, what does each side want? For America, it's pretty clear. As before, it wants to maintain its technological lead over China, reap the economic benefits, and shape the standards and guardrails around the technology to maintain its national security and protect the rights and safety of its citizens.

As far as other countries are concerned, it depends on what they think the future of AI could be. If they believe that the most powerful AI models are likely to be hoarded by the US rather than made openly available, then they have a pretty strong incentive to want an arrangement that gives them access to those models. But if they believe that suitably powerful general-purpose AI models will continue to be available on an open-source basis – I described the likelihood of a mix of open and closed general-purpose models being the norm in Chapter 9 – then the calculation becomes somewhat different. They won't necessarily want an arrangement that guarantees access to the models themselves, because they will have that anyway. But if they want those models to be trained in a way that is optimised for local languages, culture and societal needs, then the models need to be trained on local data. And they will want to be able to fine-tune these models locally, which requires high levels of technical expertise, and to use these fine-tuned models to run queries against the model and generate predictions (known as 'inference'), which requires data centres and a lot of energy. That adds up to a lot of talent and expensive hardware. Without them, countries will be left trying to make the most of models built, in large part, by American companies for American-type use.

In a world where nations are seeking ever greater sovereign control over technology, being forced to wear Uncle Sam's hand-me-downs self-evidently won't do. And if other countries want to stay at the forefront in sensitive areas like national security, defence and cybersecurity, they will want access to the specialised models – or specialised versions of open-source models – which will inevitably be tightly locked down by the US government and security agencies.

The answer is a multifaceted arrangement that pools resources and gives signatories shared access to – thereby lowering the cost of – computing power and data-centre infrastructure, data sets for training and cultural localisation, and cooperation on cutting-edge research. It would also establish global standards for the responsible development and deployment of AI, and frameworks for cooperation on security issues. Such a deal could include a host of elements, including building a commonly held AI graphics processing unit pool – essentially giving smaller nations access to the computing power needed to fine-tune their own models, develop their own AI applications and collaborate on the research and development of foundational AI technology. The deal could also include multi-stakeholder agreements between governments, the AI industry and universities; reducing barriers in trading material resources necessary for the development of AI and other advanced technologies such as quantum computing; developing a semiconductor supply chain that is independent from China and promoting the supply chain's resilience in times of crisis, expanding on the provisions of the protectionist CHIPS Act; developing common standards and benchmarks for trustworthy AI and quantum computing; cooperating on matters of AI defence and security, including intelligence sharing and the co-development of AI security technologies (in certain cases, AI-based defensive military equipment); and expanding access to the internet and AI to under-served areas of the global South.

In practice, such an arrangement would need to be tiered. It is somewhat fanciful to imagine the Trump administration being willing to enter into such a wide-ranging agreement with all comers. But deals are there to be struck over these various elements with countries that recognise it is in their interests to be part of the club. So, for example, while it is beneficial for all involved to have as many like-minded nations as possible sign up to a deal that includes open data flows and shared access to AI infrastructure, it is unrealistic to expect wide and open cooperation on sensitive areas of national security, intelligence and defence. As a starting point, cooperation on these matters would be reserved for a higher tier of close allies – just as the 'Five Eyes' alliance allows for close intelligence sharing between the US, the UK, Canada, Australia and New Zealand. For those who want access to the higher tier's sensitive collaboration, signing up to the wider deal would be an important first step.

A core component of these arrangements would be what are known as technology transfer agreements (TTAs), which are a means of transferring knowledge and technical infrastructure between two or more organisations, nationally and internationally. TTAs are developed for exactly the kind of information collaboration that is proprietary and high-risk from a national security standpoint. In the past, they have been successfully used in the semiconductor industry, for example, and today are most commonly employed in high-tech fields such as pharmaceuticals, biotechnology and nanotechnology, microelectronics, physics and the defence and security industries. Even US government entities operating with the highest standards of national security, such as the NSA, have utilised these agreements as part of their operations. And, in Europe, TTAs are formally recognised by the EU and used by organisations such as the European Defence Agency.

In developing countries, meanwhile, TTAs are seen as a powerful way to boost their economies, giving them access to technologies that would otherwise be beyond their reach. In the decades after the Second World War, Japan used TTAs as a way to grow its economy. And today, India sees TTAs as a major avenue for the development of its technology sector. The US government is currently proactively pursuing a policy of information sharing and collaboration with India in multiple tech sectors – in defence technology, advanced telecoms, and semiconductors.

The long history of TTAs between nations and other stakeholders means that a robust network of institutions, laws and professional norms already exists that could be adapted to the complexities that an omnibus US-led deal would most certainly require. And TTAs can have teeth – they can include international arbitration clauses that are actionable in dispute-resolution mechanisms. They are win–win for developing and developed nations: as history shows, they can benefit the economic growth of developing nations, while also benefiting advanced nations by lowering the costs of manufacturing and generating rents from licensing technologies.

Over time, other elements could be introduced to allow greater integration between the signatory nations, and stronger incentives for others to sign up. One area, for example, could be taxation. The way global internet companies are taxed has been hotly debated for years. The current international tax system was not designed to meet the challenges of today's global economy. In effect, we are taking rules created for an economic system driven by the manufacturing, importing and exporting of physical goods and trying to apply them to a borderless online world where tens of thousands of digital transactions take place every second. In recent years, a process being overseen by the OECD has brought together more than 130 countries, big and small, to try to establish a new system within which multi-national corporations that rise above certain designated thresholds for revenues and profit margins would be taxed a percentage of their profits, which would then be shared among countries and other jurisdictions in a way that better reflects the changing nature of the digital economy. The OECD has forecast that such a tax could raise between $17 billion and $32 billion a year.

But the OECD process has dragged on for years and has been plagued by stops and starts. It is hard to get thirteen countries to agree on something, let alone 130. A draft treaty was finally published in late 2023. But in early 2025, President Trump withdrew US support for the deal. Theoretically, the US could instead make global digital tax reform an incentive to entice countries to sign up to the wider deal. This way it wouldn't require consensus from the extensive membership list of the OECD, but only from those prepared to sign on to the overall deal – a tax-sharing deal among willing techno-democracies.

A new digital democracies alliance

A deal like this could come about in a couple of different ways. The US could take the lead and offer an open invitation to its allies to meet in one location and hammer out a broad memorandum of understanding on the key components, including the broad strokes of any economic and security collaboration. This coalition of the willing would then enter negotiations on the detail, creating a comprehensive programme that others could sign up to at a later date. The countries involved would be making a bold political statement to the rest of the world – and particularly China – about their willingness to collaborate on issues of key geopolitical importance such as AI and data flows.

But open invitations like these require years of groundwork before anyone gets round the negotiating table, and there's always the danger that the more countries are included, the greater the complexity of the agreement and the higher the likelihood of ending up with a wishy-washy compromise that satisfies no one. Alternatively, it could start smaller with just the big techno-democracies – the US, the EU, India and perhaps a small handful of others like the UK, Japan or South Korea – then grow over time to include an ever-increasing number of allies. The recently revitalised Trans-Pacific Partnership Agreement could serve as a model for how this process might take place. In fact, the TPP started as a small grouping of like-minded countries – without the US in this case – and grew only as others sought to join. Eventually, the US chose to insert itself in the TPP and made it its own, before the first Trump administration withdrew from the agreement in 2017.

Where could the deal be negotiated and how would the negotiations be overseen? Meaningful global cooperation requires institutions with clout, and the era of deglobalisation has eroded the power of international institutions like the WTO and the United Nations. It would take considerable political leadership to renew the authority and sense of purpose of these institutions. It may be better simply to create a new one that is fit for purpose. The remit of a possible new institution – let's call it the Digital Democracies Alliance – will depend on whether it will have binding powers or whether it will merely be a discussion forum. For a deal of this scope, the former is clearly preferable. It needs teeth. This also means that the new institution needs a dispute-resolution mechanism for the disagreements that will inevitably arise. But, most importantly, a new and purpose-built institution would be free from the cumbersome old rules and norms established by old institutions to serve old purposes. A new, global, democratic digital forum can write its own rules fit for twenty-first-century diplomacy and avoid the inertia of those built for the twentieth century.

If this all sounds too ambitious, we should remember that we have been in a similar position before. In 1944, as the end of the Second World War approached, delegations from the Allied nations gathered in Bretton Woods, New Hampshire. After a month of intense negotiations, they struck an agreement that became the foundation of a new global governance philosophy for the post-war era: if nations large and small ceded a degree of their own sovereignty to abide by the same global rules, they could avoid repeating the protectionism and economic catastrophes of the 1920s and 30s. Global institutions like the International Monetary Fund and the World Bank were created to promote economic growth and political stability.

Bretton Woods can serve as an inspiration, but it shouldn't serve as a model. It has been criticised for enshrining American post-war hegemony and imposing its economic policies and priorities on developing countries. Nor is it thought of warmly by the Trump administration. Vice President Vance, for example, has argued that the dollar's position as the global reserve currency – a result of the system agreed at Bretton Woods – has contributed to the decline in American manufacturing. And the Heritage Foundation's Project 2025, regarded as a blueprint for second-term Trump policies, called for the US to withdraw from the IMF and the World Bank. Nonetheless, Bretton Woods remains a demonstration that cooperation on this scale is possible. A new global deal for the digital age might be in the interests of the United States – and must be, as it is only conceivable with American political leadership – but it will work only if it empowers every nation that signs up to it. This must not be such an exclusive club that it fosters growing resentment elsewhere. That's the thing about network effects – the more of us there are participating, the more valuable it is for all of us. It's what made the open internet the greatest democratising force of the late twentieth and early twenty-first centuries. And it's why AI could be a democratising force for the decades to come.

What's in it for the US, EU and India?

Why should America spend the considerable resources and political capital required to lead such a deal? On the face of it, many of the elements of this deal might seem counter to US interests, and counter to the anti-globalist politics of the Trump administration. After all, the US is the leading AI power. It could simply lock other countries out of the AI race, forcing them to waste vast amounts of money and effort attempting – and probably failing – to build up their own AI infrastructure to the point where they can compete directly. That might set the world on a path to a more technologically divided future, but it would be one where the US has a pretty substantial head start. So, other than abstract ideas about democratic values, what is in it for the US?

Well, going it alone would come at a great cost to America as well as its allies. For a start, it would be shunning those allies and leaving them open to Chinese influence. After all, if America refuses to share its AI with Europe, India and other democracies, the window is wide open for its biggest geopolitical rival to fill the gap. America could unite the democratic world and build a bulwark against Chinese influence, or it could push its allies into Beijing's embrace. And while the US has the lion's share of the technical capacity and infrastructure right now, if the Europeans, Indians, South Koreans, Brits and Australians were forced to build up their own AI capacity, the US would be incentivising these nations and others to deny US companies access to their markets. Collaboration among allies has significant economic and political benefits for all involved, especially in such a fragmented and increasingly fragile and hostile world. And when you get down to brass tacks, there are some other pretty big incentives for the superpower leading the deal.

First, TTAs would make markets in the developing world more accessible to US companies. For example, through TTAs, the US and others will be able to license AI-related IP and sell infrastructure to companies in India and elsewhere, providing privileged access to huge markets. They could also help reduce costs by sharing some of the production of AI with countries that have lower labour and other resource costs. Indeed, there's a big incentive for the US to collaborate on both energy and talent. The US doesn't have enough of either, which is why US companies are often looking for other countries in which to build their data centres and open research labs. For those other countries, broadly speaking, they will need access to the wider talent pool and infrastructure operated by these US companies, but if that talent can be attracted to their countries, and those data centres and labs can be built within their borders, then they are also more likely to be able to exercise a greater degree of bargaining power over what capabilities the models are trained for.

Second, through these deals the US will have some influence over the future direction of the world's developing economies, as well as being able to ensure that the future of AI infrastructure is built on US technical standards. Third, this would create a technological 'bloc' that is capable of cementing the US's lead over China. Finally, the US has a self-interest to lead this coalition because the alternative – each country on its own – will make it harder for the US to access necessary resources such as cheap energy, labour and other AI-related infrastructure. In other words, it's not just the benefits of the deal that make sense – the cost of *not* making a deal should also be factored into the equation.

The long and short of it is that a deal of this nature would further US soft power, commercial interests and geopolitical leverage. In this scenario, the US remains the pre-eminent technological power, it entrenches that position in relation to its chief rival, it expands its sphere of economic influence, it strengthens its alliances, and it ensures that key values like free expression and commitment to human rights remain embedded in the architecture of the internet. US global leadership is not just good for the democratic world; it's also very good for the US.

But if the deal is that good for the US, why should the EU sign up to it? Europe is at something of a crossroads technologically. A regulatory innovator, but an innovation slacker. An enormous market of consumers, resources and talent, but a technology sector that lags far behind the US and China. If America seeks to lock down its AI infrastructure, Europe faces the prospect of spending vast amounts of money, resources and political capital on trying to keep up with the bigger boys, with little hope of actually catching them. Individual EU nation states do not have the economic capacity to develop their own cutting-edge AI foundation models or data centres on the scale required to compete. Even in the case of relatively large economies like Germany or France, any state-funded AI development programmes will pale in comparison with the resources available to the US or China.

Allocating public resources to build the specialised infrastructure needed for advanced AI is also highly inefficient, with dramatic opportunity costs as funding is diverted away from other public needs. Will European governments – already suffering productivity crises and looming deficits – really choose to build data centres instead of schools and hospitals? Will their companies be able to compete on an equal footing with the US tech giants in the market for AI chips, and if they do, do they risk the US seeking to cut them out the way they sought to cut China out?

This may be an argument that suggests European governments should step back from major investments in AI model development and focus instead on building on open-source models developed elsewhere. Yet in the age of deglobalisation and resurgent national sovereignty it is inconceivable that advanced European economies would choose not to invest in sovereign AI capacity of some kind, especially in the face of the geostrategic uncertainty created by the Trump administration's relationship with Putin's Russia, its shift in support for Ukraine, and its antagonism towards its erstwhile European allies over tech regulation, defence spending and much else. Failure to build up some kind of domestic AI capability would lock European countries into a dependence on US technology that could put them at a profound strategic disadvantage over the long term. Or, worse, they could become dependent on China instead, and in doing so weaken the bonds with their democratic cousins across the Atlantic.

In a world where cutting-edge, US-developed general-purpose models are open-sourced, there will still be important elements that European nations need: first, access to world-leading talent and the vast data-centre capacity required to train and run foundational AI models; second, access to specialised models for security, intelligence and defence purposes; and third, models trained with local data and therefore with local language, culture and social norms hardwired into them.

So they face a dilemma: attempt to compete at great economic and social cost and probably fail, or fail to compete at all. On the other hand, joining a US-led deal would be a win–win scenario. European governments could pool their resources with the US in order to get access to US AI technologies while avoiding wasting scarce resources and falling behind. And if they did that, they could focus their efforts on playing to their strengths. With its huge single market of 450 million consumers, a high-quality university sector producing top talent, its significant capacity for research and development, and its ability to set the regulatory terms, Europe could become a world leader in the application layer of AI – creating the apps and services through which people will experience it. It might even produce some of those future tech giants that European policymakers pine for.

Over time, a US-led omnibus deal would lead to a web of technical interdependence between democratic allies – the US, the EU, India and others – that would underpin the economic growth of all involved and help produce superior digital technologies. This approach ensures that all countries have a common interest in the success of their pooled resources, and that each country has at least some degree of leverage over the others by owning some of the integral parts of the AI infrastructure. An interesting parallel example of this approach is the aeroplane manufacturer Airbus, which operates as a consortium of cross-border manufacturing between France, Germany, Spain and the UK. As a key signatory to a deal of this nature, Europe's position at the top table of AI development and deployment would be assured.

If those incentives apply for the EU, they apply even more emphatically for India. With its market of more than 1.4 billion people, deep talent pool and fast-growing domestic technology sector, India too could be a leader in the application of AI, but with a scale that dwarfs the EU. Being a key signatory to an omnibus deal that keeps data flows open and gives India access to US foundational AI technology would be a win–win for India and one that is emphatically in step with the Modi government's strategy to establish the country as a global tech power with its own digital public infrastructure. The Indian government sees DPI as a way to unite the country, as well as technology that can be exported to other countries. India's tech sector is booming, India's power and prestige is growing, and India is determined to bolster its position on the global stage. It sees technology as the key to modernising its economy and sharing its prosperity in every corner of its vast country. This deal would support all of the above.

China looms large in its regional sphere of influence. Currently, India is to some degree dependent on China for its technological development, but it is actively trying to detach itself and assert independence. A deal with the US and its democratic allies could include tailored support for India's tech sector and its military defence capabilities, helping India to pivot away from its regional rival and towards its co-signatories. Indeed, the US is currently pursuing a partnership in defence technology, advanced telecoms and semiconductors with India, partly in order to create a bulwark against Chinese influence in the region. An agreement that would include data flows, AI technology transfer and collaboration between advanced economies like the US, the EU and India would probably result in huge economic benefits for India – especially because of the Indian government's interest in making AI chips.

From 2019 onwards, the Indian government flirted with a harder form of data localisation than it ultimately adopted, softening its approach largely because of the resounding message from its business sector about the importance of open data flows. Through a deal of this nature, India could boost its domestic tech sector and increase the potential for exporting its DPI.

Saving the internet

The story of the early twenty-first century is one of power being disrupted. Power has been taken away from national governments, traditional media corporations and other economic and political institutions of the pre-internet world. And it has been funnelled in two new directions: to people in general, through the democratising effect of the internet, social media and, increasingly, generative AI; and to the small number of Big Tech giants. The combination of the techlash, the reversal of globalisation and the emergence of AI has brought the world to a fork in the road. The pendulum has swung from utopianism to doomerism as debates have raged about the impact of this technological disruption on our democracies, our agency and the welfare of our children. The status quo is undesirable, but the likely alternative, if we are not careful, could be far worse. We need a new balance of power – between people and platforms; between Big Tech and nation states; and between the societies that make up the democratic world.

Earlier in this book, I posed the question: what do we want the internet to be? I think most reasonable people would agree that the ability to communicate, share information and ideas, buy and sell products and services, and otherwise interact and collaborate openly across borders has been not just good economically, but something we all value in our daily lives. We want to maintain our ability to do these things, but to do them safely, in a way that protects our privacy over things we believe should be kept private, and in a way that ensures we're in control of our interactions and not being unduly manipulated or exploited. We want new technologies to keep coming when they make our lives easier or more enjoyable, but we want the risks to be considered, managed and mitigated. We want the people building the technologies to be accountable. We want our concerns to be listened to and acted upon. And we don't want politics or commerce to rob us of the benefits that technologies bring us. Fundamentally, we want technology to serve us as individuals and as societies. All of these things are possible. But they are not inevitable.

It may seem counter-intuitive, perhaps even foolhardy, to argue for such an emphatic return to international cooperation in an era when all the global trends are pointing in the opposite direction, just as it may seem unrealistic to expect big technology companies to open themselves up willingly to greater scrutiny and accountability. If these ambitions are too strongly opposed by the prevailing political climate, then we may well be stuck on our current trajectory. The internet will probably splinter into several local and regional internets, run according to different rules and standards. Government censorship and surveillance will increasingly become the norm, in part because partitioning the open internet will inevitably weaken the democratic values that were hardwired into its open structure, and in part because the temptation to stifle dissent and snoop on citizens will be too great, even in the most robust of democracies. I saw at first hand during my own time in government how easily ministers can fall under the spell of the securocrats, and I've seen the same tendency to a greater or lesser degree in the politicians I've interacted with on Meta's behalf in the US, the UK, India, Brazil and other countries across Europe, Latin America and Asia. If governments can turn the infrastructure of the internet to their political advantage, they surely will. The only thing that can restrain this impulse is the even greater pull of economic progress and geopolitical self-interest.

The world's staggering progress in living standards over the last two centuries or so owes a great deal to both technological progress and open trade. And history has shown us time and time again that economic cooperation between nations is the surest path to peace and stability. At a historical moment when globalisation is unwinding, geopolitical tensions are rising and war has reared its ugly head once again in eastern Europe and the Middle East, technological development brings us to that critical fork in the road. It is up to us to choose cooperation or fragmentation.

Without a deal on AI between democratic allies, the AI race forces political dilemmas on everyone involved. Should America go it alone and risk pushing its allies towards China? Should Europe or India spend vast amounts of money and energy building their own AI infrastructure and still risk falling behind? There are rarely easy choices in politics. But one path creates a technological alliance among the key players in the democratic world, while the other leads to a grim, technologically divided future. I don't think the latter has to be our fate. Yes, these proposals are optimistic, but I don't believe they are naive or foolishly utopian. Like everything else in this book, they are driven by the realities of power.

For tech companies, letting the sunlight in is, in the end, inevitable. Doing it voluntarily is the best way for them to keep doing what they really want to do: build new things and be rewarded for doing so. For national governments, ambitious as this proposed deal is, it is driven fundamentally by realpolitik. If the era of deglobalisation has achieved anything at a geopolitical level, it has forced governments to make hard choices about where their priorities lie. This deal would be in the economic and political interests of the world's techno-democracies – both individually and collectively. As the supreme technological power, it is in the United States' geostrategic interest to lead the deal, reinforcing its position against its greatest rival, bolstering its economic interests and furthering its soft power. And it is in the interests of Europe, India and other techno-democracies that want to boost their domestic tech sectors and economies, have influence over and access to vital AI infrastructure without entering a costly and probably doomed 'arms race', and spread the benefits of these technologies throughout their own societies. You never know, it might even renew the waning appetite for global cooperation and collaboration more broadly, helping us to meet other global challenges, not least on climate change. At the very least, these proposals are in keeping with the fundamental spirit of the internet – open, collaborative and democratic.

We've seen that we can't turn the clock back and reinstate the world as it was before the internet, social media or AI. And we shouldn't want to. These things are fundamentally liberating technologies that give billions of people more agency in their lives. But the world is changing rapidly, and things are still in flux. It's not too late to establish a new balance of power and build a new consensus about what we want the internet to be. It will take openness from politicians and technologists alike. Openness is, after all, what built the internet. It was there in the ethos of the ARPANET scientists. It was there in the early idealism of the Silicon Valley pioneers. It was hardwired into the internet's design and at the heart of its remarkable, revolutionary success. And openness remains our best hope of saving the internet today. It won't be easy, and it certainly isn't inevitable.

But it's possible.

Acknowledgements

When you write a book about politics and technology, it is hard to avoid giving the impression that every insight is yours alone. While it would be flattering to be thought of as what Americans rather grandly refer to as a 'thought leader', the truth is that this book and every experience it has been built on has been a collaboration. For all the challenges, dilemmas and stresses that working in both politics and technology present, the great joy of both worlds is that they are team pursuits. In every phase of my career I've been able to rely on the support, advice and expertise of colleagues, staff, peers, outsiders and sometimes even rivals. And as a refugee from the world of British politics into the world of commerce and tech in Silicon Valley, I have relied on a huge number of people to educate and guide me – far too many to be able to list in full here.

That said, there are some people to whom I owe a specific debt of gratitude. First and foremost, this book wouldn't exist without Phil Reilly. We have written a book together before, but on a wholly different subject – the fact that he was able to master the material for this one with the same versatility and fluid pen as before is a tribute to his immense talents as a writer and synthesiser of complex ideas. It would also not exist if it were not for the diligence of Andreas Katsanevas, whose deep research provided the backbone for so many of the arguments this book contains.

Will Hammond is everything an author could want from an editor – thoughtful and wise at all times, but precise and demanding when required. This book is much improved for his expert eye. I want to thank him and all the team at The Bodley Head for the trust they placed in me and their willingness to take on a book that puts forward a nuanced – and therefore unfashionable – case about the role of technology in society.

I'm particularly grateful for the support of Tim Colbourne, Elaine Sedenberg, Lena Pietsch, Jimmy Raimo and Jennifer Broxmeyer, who were an invaluable sounding board for me as I developed the arguments in the book and whose ideas and insights have been absorbed into the text. I am also grateful for the expertise of a number of my colleagues at Meta whose feedback on early drafts was invaluable, most especially Andrew 'Boz' Bosworth, Chris Cox, Rob Sherman, Monika Bickert, Joel Kaplan, David Ginsberg, Chris Yiu, Andrew Hall and Rachel Lieber.

I have had the great pleasure of being able to discuss my ideas with a number of people with deep expertise on matters of technology and society. Again, there are too many brains I have picked to mention, but I would like to thank Richard Allan, Verity Harding, Yann LeCun, Alex Schultz and Nate Persily in particular, all of whom have influenced my thinking on a range of important matters. This book also relied on the work of a range of people across academia, journalism and business whose books, articles and reports I have cited. I hope they feel I have reflected their work fairly.

And, of course, I owe a particular debt to Mark Zuckerberg and Sheryl Sandberg, who gave me the opportunity to see the world of technology from the inside and put a significant amount of trust in me despite being an outsider to their world. They are both remarkable leaders, from whom I have learned an immense amount.

Notes

Prologue

p. 1, A hundred million people were using ChatGPT: Krystal Hu, 'ChatGPT sets record for fastest-growing user base – analyst note', Reuters, 2 February 2023

p. 3, We're going to work with President Trump: Mark Zuckerberg, 'More Speech and Fewer Mistakes', Meta website, https://about.fb.com/news/2025/01/meta-more-speech-fewer-mistakes/, 7 Janu-ary 2025

Introduction: The Corridors and Campuses of Power

p. 11, *A Declaration of the Independence of Cyberspace*: Electronic Frontier Foundation website, 8 February 1996, https://www.eff.org/cyberspace-independence

p. 14, the archetypal villain of movies: Jake Coyle and AP, 'Tech bros are Hollywood's latest super villains', *Fortune*, 7 March 2023

p. 17, was called Tempora: Ewen MacAskill et al., 'GCHQ taps fibre-optic cables for secret access to world's communications', *Guardian*, 21 June 2013

p. 19, Internet Research Agency bot farm: 'UK exposes sick Russian troll factory plaguing social media with Kremlin propaganda', press release, UK Government website, 1 May 2022

p. 19, later debunked by the UK's Information Commissioner's Office: ICO letter of 2 October 2020 to Julian Knight MP, ICO website

p. 24, disclosed to investors as a potential business risk: Alex Weprin, 'Meta warns that Mark Zuckerberg's love of MMA could hit its bottom line', *Hollywood Reporter*, 5 February 2024

1 *The Trouble with Social Media*

p. 33, to 'distract, divide, and madden': Jason Pontin, 'The case for less speech', *Wired*, 6 November 2018

p. 33, YouTube 'may be one of the most powerful radicalizing instruments . . .': Zeynep Tufekci, 'YouTube, the great radicalizer', *New York Times*, 10 March 2018

p. 33, And in a 2022 article in *The Atlantic*: Jonathan Haidt, 'Why the past 10 years of American life have been uniquely stupid', *The Atlantic*, 11 April 2022

p. 35, media scholar Ella Hafermalz notes: Ella Hafermalz, 'Book review – *The Age of Surveillance Capitalism*', *M@n@gement*, vol. 24, 2021, issue 4

p. 36, 'hijacked our minds': Nicholas Thompson, 'Our minds have been hijacked by our phones. Tristan Harris wants to rescue them', *Wired*, 26 July 2017

p. 37, 'Tristan Harris's argument': Ibid.

p. 38, more than a thousand advertisers joined a boycott of Facebook: Hongwei He et al., 'What can we learn from #StopHateForProfit boycott regarding corporate social irresponsibility and corporate social responsibility?', *Journal of Business Research*, vol. 131, July 2021, pp. 217–26

p. 38, big-brand advertisers abandoned Twitter/X: 'Advertisers leave Elon Musk's X after their ads appeared next to pro-Nazi posts', transcript, NPR, 22 November 2023

p. 40, In an article I wrote in 2021: Nick Clegg, 'You and the algorithm: it takes two to tango', Medium, 31 March 2021

p. 43, Research from Stanford: Levi Boxell et al., 'Cross-country trends in affective polarization', Stanford University, August 2021

p. 43, an Oxford University literature review: Amy Ross Arguedas et al., 'Echo chambers, filter bubbles, and polarisation: a literature review', Reuters Institute, 2022

p. 44, levels of ideological polarisation are similar: An Nguyen and Hong Tien Vu, 'Testing popular news discourse on the "echo chamber" effect: does political polarisation occur among those relying on social media as their primary politics news source?', *First Monday*, vol. 24, no. 6, 3 June 2019

p. 44, A Harvard study: Yochai Benkler et al., 'Mail-in voter fraud: anatomy of a disinformation campaign', Berkman Klein Center Research Publication no. 2020-6, 2 October 2020

p. 44, Reuters Institute in 2017: Nic Newman et al., *Reuters Institute Digital News Report 2017*

p. 45, Pew in 2019: Laura Silver and Christine Huang, 'In emerging economies, smartphone and social media users have broader social networks', Pew Research Center, 22 August 2019

p. 45, the scientific journal *Nature*: Brendan Nyhan et al., 'Like-minded sources on Facebook are prevalent but not polarizing', *Nature*, 27 July 2023

p. 46, the co-chairs of the study: 'EXPERT REACTION: Facebook and Instagram "echo chambers" may not be driving our political polarization', Scimex.org, 28 July 2023

p. 46, leaving Facebook for a month reduced political polarisation: Hunt Allcott et al., 'The welfare effects of social media', *American Economic Review*, vol. 110, no. 3, March 2020, pp. 629–76

p. 46, exposure to opposing views: Christopher A. Bail et al., 'Exposure to opposing views on social media can increase political polarization', *PNAS*, vol. 115, no. 37, pp. 9216–21, 28 August 2018

p. 46, Joshua Tucker and Stanford scholar Nate Persily: Nathaniel Persily and Joshua A. Tucker (eds), *Social Media and Democracy: The State of the Field and Prospects for Reform*, SSRC Anxieties of Democracy series, Cambridge University Press, 2020

p. 48, *The Anxious Generation*, in which he makes the case: Jonathan Haidt, *The Anxious Generation: How the Great Rewiring of Childhood Is Causing an Epidemic of Mental Illness*, Penguin Press, 2024

p. 48, An international study which looked at more than 900,000 adolescents: Alina Cosma et al., 'Cross-national time trends in adolescent mental well-being from 2002 to 2018 and the explanatory role of schoolwork pressure', *Journal of Adolescent Health*, vol. 66, issue 6, supplement, June 2020, pp. S50–S58

p. 48, Max Roser, of Our World in Data: X post, https://x.com/MaxCRoser/status/1641146860281659406, 29 March 2023

p. 48, World Health Organization data: 'Reported suicide rates among young people', Our World in Data; data source: WHO Mortality Database (2024)

p. 49, recorded rates of teenage suicide and depression have increased: Adriana Corredor-Waldron and Janet Currie, 'To what extent are trends in teen mental health driven by changes in reporting? The example of suicide-related hospital visits', National Bureau of Economic Research, working paper 31493, July 2023

p. 49, contributed an article to *Nature*: Candice L. Odgers, 'The great rewiring: is social media really behind an epidemic of teenage mental illness?', *Nature*, 29 March 2024

p. 49, an article in *The Atlantic*: Candice L. Odgers, 'The panic over smartphones doesn't help teens', *The Atlantic*, 21 May 2024

p. 49, David Wallace-Wells: David Wallace-Wells, 'Are smartphones driving our teens to depression?', *New York Times*, 1 May 2024

p. 50, the *Wall Street Journal* declared: Georgia Wells et al., 'Facebook knows Instagram is toxic for teen girls, company documents show', *Wall Street Journal*, 14 September 2021

p. 51, danah boyd has argued: danah boyd, 'Struggling with a moral panic once again', Data: Made Not Found (by danah), 18 April 2024

p. 51, Experimental psychologist Amy Orben: Amy Orben et al., 'Windows of developmental sensitivity to social media', *Nature Communications*, vol. 13, article no. 1649 (2022)

p. 53, psychologist Jacqueline Nesi: Jacqueline Nesi, 'Making sense of The Anxious Generation', Techno Sapiens, 29 April 2024

p. 54, average teen uses forty different apps: Jenny S. Radesky et al., 'Constant companion: a week in the life of a young person's smartphone use', Common Sense, 2023

2 *The Techlash*

p. 57, 'Lunacy in England': *New York Times* website, https://timesmachine. nytimes.com/timesmachine/1894/08/12/106870661.pdf

p. 57, a condition called 'bicycle face': Janet M. Davis, *The Circus Age: Culture & Society Under the American Big Top*, University of North Carolina Press, 2002, p. 91

p. 57, 'a plunging kind of motion': Beata Kiersnowska, 'Female cycling and the discourse of moral panic in late Victorian Britain', *Atlantis*, vol. 41, no. 2, 2019

p. 58, 'telephone ear': *Buffalo Evening News*, 29 June 1896, p. 35

p. 58, 'telephone mania': *New-York Tribune*, 4 April 1897, p. 10

p. 58, 'calculated to upset the nervous system': *The Times-Democrat*, 12 March 1882, p. 4

p. 58, spreading indecent and obscene images: Pessimists Archive website, https://pessimistsarchive.org/list/photography/clippings

p. 58, women and children vulnerable to predators: Ben Rooney, 'Women and children first: technology and moral panic', *Wall Street Journal*, 11 July 2011

p. 58, corrupting young minds: Pessimists Archive website, https://pessimists archive.org/list/novels/clippings

p. 58, 'modern Frankenstein' and a 'suicidal folly': *Kearney Daily Hub*, 13 October 1938, p. 1

p. 58, rioting as early as 1881: newspaper report of 26 February 1881, Pessimists Archive website, https://pessimistsarchive.org/list/vaccines/clippings/1881/m-sc-66-421

p. 58, 'destroying the instinct of motherhood': *Davenport Weekly Democrat and Leader*, 11 July 1907, p. 10

p. 58, 'much as a chronic alcoholic does drink': Amy Orben, 'The Sisyphean cycle of technology panics', *Perspectives on Psychological Science*, vol. 15, issue 5, 30 June 2020

p. 58, violence on television: Christopher J. Ferguson and Cathy Faye, 'A history of panic over entertainment technology', *Behavioral Scientist*, 1 January 2018

p. 58, 'as addictive as heroin': Lee Price, 'Playing games as addictive as heroin', *Sun*, 8 July 2014

p. 59, a direct line to drug dealers: Louis Anslow, 'The forgotten war on beepers', Pessimists Archive, 10 April 2024

p. 61, 'all the furious fence-building': Anne O'Hare McCormick, 'The radio: a great unknown force', *New York Times Magazine*, 27 March 1932

p. 62, The *New York Times* howled: 'Radio listeners in panic, taking war drama as fact', *New York Times*, 31 October 1938

p. 62, CBS's Frank Stanton recalled later: Corydon B. Dunham, *Fighting for the First Amendment: Stanton of CBS vs. Congress and the Nixon White House*, Praeger, 1997

p. 62, in *Slate* magazine: Jefferson Pooley and Michael J. Socolow, 'The myth of the *War of the Worlds* panic', *Slate*, 28 October 2013

p. 62, W. Joseph Campbell wrote in 2010: W. Joseph Campbell, *Getting It Wrong: Debunking the Greatest Myths in American Journalism*, Second Edition, University of California Press, 2016

p. 64, According to the Pew Research Center: Newspapers fact sheet, Pew Research Center, 10 November 2023

p. 64, favoured candidate to be the chair of Ofcom: Laura Kuenssberg, 'Paul Dacre and Ofcom: what's going on?', BBC, 28 May 2021

p. 64, twice nominated for a peerage by Boris Johnson: Kiran Stacey, 'Boris Johnson nominates *Daily Mail* chief Paul Dacre for peerage for second time', *Guardian*, 9 March 2023

p. 66, as Tim Berners-Lee ... warned: Rod McGuirk and Kelvin Chan, 'Australian media law raises questions about "pay for clicks"', AP News, 18 February 2018

p. 67, science and technology professor Lee Vinsel: Lee Vinsel, 'You're doing it wrong: notes on criticism and technology hype', Medium, 1 February 2021

p. 67, 'what historian David C. Brock calls "wishful worries"': David C. Brock, 'Our censors, ourselves: commercial content moderation', *Los Angeles Review of Books*, 25 July 2019

p. 68, In an essay for *The Atlantic*: Jonathan Haidt, op. cit.; see p. 33 above

p. 71, a large Oxford Internet Institute study: Lisa-Maria Neudert et al., 'Global attitudes towards AI, machine learning & automated decision making', Oxford Internet Institute, 7 October 2020

p. 71, Another Pew Research study: Courtney Johnson and Alec Tyson, 'People globally offer mixed views of the impact of artificial intelligence, job automation on society', Pew Research Center, 15 December 2020

p. 71, Citibank estimates: 'Asia as a time machine to the future', Citi, 25 May 2023

p. 71, Asia accounted for 52 per cent of global growth: Jonathan Woetzel and Jeongmin Seong, 'What is driving Asia's technological rise?', *Project Syndicate*, 23 December 2020

p. 72, a full-size robot replica of himself: Atsuo Yamaguchi, 'Robo-Kono: researchers unveil robotic avatar of Japan's digital minister', *Mainichi*, 22 October 2022

p. 72, Nirit Weiss-Blatt ... traces the eruption: Nirit Weiss-Blatt, 'Donald Trump caused the techlash', Techdirt, 14 April 2021

p. 74, Martin Gurri: Sean Illing, 'The elites have failed', *Vox*, 27 March 2021

3 *The Open Internet: How Big Tech Got Big*

p. 80, *From Counterculture to Cyberculture*: Fred Turner, *From Counterculture to Cyberculture: Stewart Brand, the Whole Earth Network, and the Rise of Digital Utopianism*, University of Chicago Press, 2006

p. 80, her fascinating book *AI Needs You*: Verity Harding, *AI Needs You: How We Can Change AI's Future and Save Our Own*, Princeton University Press, 2024, pp. 136–7

p. 81, Steve Wilhite: Harry McCracken, 'How Steve Wilhite created GIF, the graphics format that ate the world', Fast Company, 25 March 2022

p. 81, Email, for example, has a history: Craig Partridge, 'The technical development of internet email', *IEEE Annals of the History of Computing*, vol. 30, issue 2, April–June 2008

p. 84, 'aggregation theory': Ben Thompson, 'Aggregation theory', Stratechery, 21 July 2015

p. 87, Rupert Murdoch's ill-fated purchase: Dominic Rushe, 'Myspace sold for $35m in spectacular fall from $12bn heyday', *Guardian*, 30 June 2011

p. 88, In *AI Needs You*: Verity Harding, op. cit., pp. 71–120; see p. 80 above

p. 88, Alabama state Supreme Court's controversial ruling: Sabrina Talukder, 'How the Alabama IVF ruling is connected to upcoming Supreme Court cases on abortion', Center for American Progress, 11 March 2024

p. 90, 'nailing Jell-O to a wall': Lorand Laskai, 'Chapter 6 – "Nailing Jello to a Wall"', The China Story, Australian Centre on China in the World, 2016

p. 91, from 2 per cent of China's population using the internet: World Bank Group data; World Bank website, https://data.worldbank.org/indicator/IT.NET.USER.ZS?end=2023&locations=CN&start=1990&view=chart

p. 91, There are now more than sixty state agencies: Alina Polyakova and Chris Meserole, 'Exporting digital authoritarianism: the Russian and Chinese models', Foreign Policy at Brookings, Brookings Institution, August 2019

p. 91, in a report for the Brookings Institution: Ibid.

p. 92, Lorand Laskai notes: Lorand Laskai, op. cit.; see p. 90 above

4 *Nothing More Controversial Than Speech*

p. 95, 'liberation technology': Larry Diamond and Marc F. Plattner (eds), *Liberation Technology: Social Media and the Struggle for Democracy*, Johns Hopkins University Press, 2012

p. 102, has a specific history: Jackie Flynn Mogensen, 'Why Trump's "looting" tweet was even worse than you thought', Mother Jones, 29 May 2020

p. 103, in a Facebook post: Mark Zuckerberg, Facebook post, 30 May 2020

p. 107, a video of then House Speaker Nancy Pelosi: Jim Waterson, 'Facebook refuses to delete fake Pelosi video spread by Trump supporters', *Guardian*, 24 May 2019

p. 108, may have known the attacker: Oliver Darcy and Donie O'Sullivan, 'Elon Musk, Twitter's new owner, tweets conspiracy theory about attack on Paul Pelosi', CNN Business, 31 October 2022

p. 108, was a 'pedo guy': 'Cave diver tells court Elon Musk tweets "humiliated" him', BBC, 5 December 2019

p. 108, firing employees who criticised him: Kate Conger et al., 'Elon Musk fires Twitter employees who criticized him', *New York Times*, 15 November 2022

p. 110, mass protests in Iran: Raksha Kumar, 'Not quite the Arab Spring: how protestors in Iran are using social media in innovative ways', Reuters Institute, 6 December 2022

p. 110, restricted the use of the internet: Yasmin Green, 'Iran's internet blackouts are part of a global menace', *Wired*, 19 October 2022

p. 110, apps like Instagram: 'As unrest grows, Iran restricts access to Instagram, WhatsApp', Reuters, 21 September 2022

5 *How Big a Deal Is Generative AI?*

p. 118, rules to address the issues raised by generative systems: Ryan Morrison, 'European lawmakers vote to adopt EU AI Act', Tech Monitor, 14 June 2023

p. 118, cure cancer: Marc Emmer, 'Could AI help cure cancer?', Inc., 23 October 2023

p. 118, harness nuclear fusion: Angela Dewan, 'Scientists say they can use AI to solve a key problem in the quest for near-limitless clean energy', CNN, 21 February 2024

p. 118, solve the climate crisis: Amit Katwala, 'DeepMind wants to use AI to solve the climate crisis', *Wired*, 18 October 2023

p. 118, live on Mars: Vishwam Sankaran, 'AI can help find caves on Mars for future astronauts to live in, scientists say', *Independent*, 18 January 2024

p. 119, new advanced therapies: Mary Kekatos, 'How artificial intelligence is being used to detect, treat cancer – and the potential risks for patients', ABC News, 21 July 2023

p. 119, diagnoses for diseases: Bo Zhang et al., 'Machine learning and AI in cancer prognosis, prediction, and treatment selection: a critical approach', *Journal of Multidisciplinary Healthcare*, vol. 16, pp. 1779–91, 26 June 2023

p. 119, like cancer: Universitat Politècnica de València, 'New artificial intelligence algorithms facilitate diagnosis of difficult cancers', News Medical Life Sciences, 27 November 2023

p. 119, and diabetes: 'New AI technology shows promise in early detection of diabetes using X-rays and medical records', UTMB News, 20 July 2023

p. 119, the discovery of antibiotics: Gary Liu et al., 'Deep learning-guided discovery of an antibiotic targeting *Acinetobacter baumannii*', *Nature Chemical Biology*, vol. 19, pp. 1342–50, 25 May 2023

p. 119, accurate and detailed MRIs: 'Accelerating MRI reconstruction via active acquisition', Meta, 18 June 2019

p. 119, aviation carbon emissions: Carl Elkin and Dinesh Sanekommu, 'How AI is helping airlines mitigate the climate impact of contrails', Google blog, 8 August 2023

p. 119, predict weather patterns: Aaron Frank, 'NVIDIA is making a digital twin of earth's climate', Singularity Hub, 4 October 2022

p. 119, the protein folding problem: Aisha Al-Janabi, 'Has DeepMind's AlphaFold solved the protein folding problem?', *BioTechniques*, vol. 72, issue 3, 4 February 2022

p. 119, first ever images of unseen parts of the sun: Ian Randall, 'Artificial intelligence reveals first-ever image of unseen parts of the sun', *Newsweek*, 16 November 2023

p. 119, decipher ancient scrolls: Nuray Bulbul, 'Researchers use AI to decipher word on ancient scroll charred by Mount Vesuvius', *Standard*, 13 October 2023

p. 119, as Natasha Loder ... notes: 'Simply Science' email newsletter, *The Economist*, 27 March 2024

p. 119, Harris declared: Steven Levy, 'How to start an AI panic', *Wired*, 10 March 2023

p. 120, to use AI to speak to whales: Billy Perrigo, 'Aza Raskin tried to fix social media. Now he wants to use AI to talk to animals', *Time*, 11 December 2022

p. 121, An open letter: Cade Metz and Gregory Schmidt, 'Elon Musk and others call for pause on AI, citing "profound risks to society"', *New York Times*, 29 March 2023

p. 121, 'The AI doomers' playbook': Nirit Weiss-Blatt, 'The AI doomers' playbook', Techdirt, 14 April 2023

p. 121, regimes using AI for cyberattacks: Tad Friend, 'Sam Altman's Manifest Destiny', *New Yorker*, 10 October 2016

p. 121, Ben Thompson has a theory: Ben Thompson, 'Attenuating innovation (AI)', Stratechery, 1 November 2023

p. 121, telling the *Australian Financial Review*: John Davidson, 'Google Brain founder says big tech is lying about AI extinction danger', *Australian Financial Review*, 30 October 2023

p. 123, As the *New York Times* points out: Tiffany Hsu and Stuart A. Thompson, 'Disinformation researchers raise alarms about AI chatbots', *New York Times*, 20 June 2023

p. 125, to triple from 2022 levels by the end of the decade: Vivian Lee, 'The impact of GenAI on electricity: how GenAI is fueling the data center boom in the US', LinkedIn, 13 September 2023

p. 125, Snapdragon and Google: Richard Waters, 'The race to bring generative AI to mobile devices', *Financial Times*, 16 May 2023

p. 125, GPT-4 took around $100 million to train: 'Simply Science' email newsletter, *The Economist*, 17 April 2024

p. 126, Meta's I-JEPA AI model: 'I-JEPA: the first AI model based on Yann LeCun's vision for more human-like AI', Meta, 13 June 2023

p. 128, As Bill Gates describes: Bill Gates, 'AI is about to completely change how you use computers', Gates Notes, 9 November 2023

p. 131, She argues that generative AI: Cami Rosso, 'New study highlights opportunities for artificial emotional intelligence', *Psychology Today*, 24 August 2022

p. 131, In an article for NPR: Yuki Noguchi, 'Therapy by chatbot? The promise and challenges in using AI for mental health', NPR, 19 January 2023

p. 134, *Atlantic* reporter Ethan Brooks: Ethan Brooks, 'You can't truly be friends with an AI', *The Atlantic*, 14 December 2023

p. 136, we learn how to empathise with others: Ashley Abramson, 'Cultivating empathy', *Monitor on Psychology* (American Psychological Association), vol. 52, no. 8, 1 November 2021

p. 136, the Paperclip Problem: Nick Bostrom, 'Ethical issues in advanced Artificial Intelligence', in Susan Schneider (ed.), *Science Fiction and Philosophy: From Time Travel to Superintelligence*, Wiley-Blackwell, 2009

p. 136, 'We need to control AI agents now': Jonathan L. Zittrain, 'We need to control AI agents now', *The Atlantic*, 2 July 2024

p. 137, Vannevar Bush: Vannevar Bush, 'As we may think', *The Atlantic*, July 1945

p. 137, ENIAC: 'ENIAC display', University of Michigan Computer Science and Engineering website, https://cse.engin.umich.edu/about/history/eniac-display/

p. 137, inventions like the mouse: 'The mouse', Computer History Museum website, https://www.computerhistory.org/revolution/input-output/14/350

p. 138, 1980s and up to the early 1990s: Jonathan Weber, 'Cult. Company. Chaos: Apple computers had it all first . . .', *Los Angeles Times*, 10 December 1995

p. 139, dating apps: Amanda Hoover, 'We tried a dating app that lets a chatbot break the ice for you. It got weird', *Wired*, 25 January 2024

6 The AI Power Paradox

p. 145, Commission on Inequality in Education: Nick Clegg et al., Commission on Inequality in Education report, Social Market Foundation, June 2017

p. 145, OECD's annual Education at a Glance report: 'Education at a Glance 2024', OECD, 10 September 2024

p. 147, a 2022 study by PwC: 'What does virtual reality and the metaverse mean for training?', PwC, 15 September 2022

p. 147, an earlier PwC study: 'The effectiveness of virtual reality soft skills training in the enterprise', public report, PwC, 25 June 2020

p. 147, a more recent survey: 'Insights from teachers on the future of XR for education', XR Association, December 2022

p. 147, *Experience on Demand*: Jeremy Bailenson, *Experience on Demand: What Virtual Reality Is, How It Works, and What It Can Do*, W. W. Norton & Co., 2018

p. 149, A recent study comparing the use of VR: Tzu-Yu Tai and Howard Hao-Jan Chen, 'The impact of immersive virtual reality on EFL learners' listening comprehension', *Journal of Educational Computing Research*, vol. 59, issue 7, 11 February 2021

p. 149, As far back as 2001: Maria T. Schultheis and Albert A. Rizzo, 'The application of virtual reality technology in rehabilitation', *Rehabilitation Psychology*, vol. 46, no. 3, pp. 296–311, 2001

p. 149, 'improved emotion recognition . . .': Nyaz Didehbani et al., 'Virtual reality social cognition training for children with high functioning autism', *Computers in Human Behavior*, vol. 62, pp. 703–11, September 2016

p. 149, 'a positive change in participant skills . . .': Vijay Ravindran et al., 'Virtual reality support for joint attention using the Floreo joint attention module: usability and feasibility pilot study', *JMIR Pediatrics and Parenting*, vol. 2, issue 2, 30 September 2019

p. 149, UNESCO estimated: '250 million children out-of-school: what you need to know about UNESCO's latest education data', UNESCO, 19 September 2023, updated 21 September 2023

p. 149, Smartphone usage is growing rapidly: 'Number of smartphone subscriptions in sub-Saharan Africa from 2012 to 2019' (chart), Statista, 8 October 2024

p. 150, In 1900, just 4 per cent of the US population: Loraine A. West et al., '65+ in the United States: 2010', United States Census Bureau, June 2014

p. 151, By 2020, more than a quarter of the Japanese population: 'How will population ageing affect health expenditure trends in Japan and what are the implications if people age in good health?', World Health Organization, 2020

p. 152, Most Americans, for example, believe that in 2050: Andrew Daniller, 'Americans take a dim view of the nation's future, look more positively at the past', Pew Research Center, 24 April 2023

p. 153, like Google's Gemini: Gemini website, https://gemini.google.com

p. 153, software engineers can code up to twice as fast: Eirini Kalliamvakou, 'Research: quantifying GitHub Copilot's impact on developer productivity and happiness', GitHub Blog, 7 September 2022, updated 21 May 2024

p. 153, Goldman Sachs has estimated: Jared Cohen et al., 'The generative world order: AI, geopolitics, and power', Goldman Sachs, 14 December 2023

p. 153, PwC puts that figure at $6.6 trillion: 'PwC's global artificial intelligence study: sizing the prize', PwC, 2017

p. 154, AI could affect nearly 40 per cent of all jobs: Annabelle Liang, 'AI to hit 40% of jobs and worsen inequality, IMF says', BBC, 15 January 2024

p. 155, As Goldman Sachs analysts: Jared Cohen et al., op. cit.; see p. 153 above

p. 156, MIT associate professor Danielle Li: Brian Eastwood, 'Workers with less experience gain the most from generative AI', MIT Sloan School of Management, 26 June 2023

p. 156, studied the use of an AI-powered conversational assistant: Erik Brynjolfsson et al., 'Generative AI at work', National Bureau of Economic Research, working paper 31161, April 2023, revised November 2023

p. 157, UCL professor Oriel Sullivan: Oriel Sullivan, 'Gender inequality in work–family balance', *Nature Human Behaviour*, vol. 3, pp. 201–3, March 2019

p. 157, A study by Canadian researchers: Martha MacDonald et al., 'Taking its toll: the influence of paid and unpaid work on women's well-being', *Feminist Economics*, vol. 11, issue 1, 2005

p. 157, A 2023 study of middle-class women and men: Natalia Reich-Stiebert et al., 'Gendered mental labor: a systematic literature review on the cognitive dimension of unpaid work within the household and childcare', *Sex Roles*, vol. 88, 29 April 2023

p. 157, Pew Research analysis: Katherine Schaeffer, 'Among US couples, women do more cooking and grocery shopping than men', Pew Research Center, 24 September 2019

p. 158, A recent OECD co-authored study claimed AI: Clementine Collett et al., 'The effects of AI on the working lives of women', UNESCO, OECD, IDB, 2022

p. 158, As Daniel Björkegren of Columbia University has argued: 'Could AI transform life in developing countries?', *The Economist*, 25 January 2024

p. 161, author and tech entrepreneur Nicolas Colin: Nicolas Colin, 'Hedge: inventing a new safety net', Medium, 7 January 2019

p. 161, 'the difficulty of finding housing': Kim-Mai Cutler, 'How burrowing owls lead to vomiting anarchists (or SF's housing crisis explained)', TechCrunch, 14 April 2014

p. 161, 'And we need new unions': Noah Smith, 'Stronger labor unions could do a lot of good', Bloomberg, 6 December 2017

7 Digital Sovereignty

p. 169, Trump threatened to jail Mark Zuckerberg for life: 'Donald Trump threatens to imprison Mark Zuckerberg "for rest of his life" if "he does anything illegal" over election', Sky News, 30 August 2024

p. 169, Vance has called for Google to be broken up: Lauren Feiner and Alex Heath, 'J. D. Vance is anti-Big Tech, pro-crypto', The Verge, 16 July 2024

p. 172, 'nerd harder': Benedict Evans, 'When tech says "no"', Benedict Evans online newsletter, 24 August 2023

p. 172, declared Morris Chang: Dashveenjit Kaur, 'Globalization and free trade is "almost dead", says TSMC CEO', Tech Wire Asia, 19 December 2022

p. 173, Pinelopi K. Goldberg and Tristan Reed: Pinelopi K. Goldberg and Tristan Reed, 'Is the global economy deglobalizing? And if so, why? And what is next?', National Bureau of Economic Research, working paper 31115, April 2023

p. 175, The Tony Blair Institute for Global Change: Jordan Kyle and Limor Gultchin, 'Populism in power around the world', Tony Blair Institute for Global Change, 13 November 2018

p. 176, A leader column in *The Economist*: 'The liberal international order is slowly coming apart', leader, *The Economist*, 11 May 2024

p. 177, 'a dangerous spiral into protectionism worldwide': 'The destructive new logic that threatens globalisation', leader, *The Economist*, 12 January 2023

p. 178, The European Centre for International Political Economy: Matthias Bauer and Dyuti Pandya, 'EU autonomy, the Brussels effect, and the rise of global economic protectionism', ECIPE, February 2024

p. 179, as Benedict Evans explains: Benedict Evans, 'Regulating technology', Benedict Evans online newsletter, 23 July 2020

p. 180, attempts to ban TikTok: Kari Paul, 'Senate passes bill banning TikTok if parent company does not sell it', *Guardian*, 24 April 2024

p. 180, the American Innovation and Choice Online Act: 'Klobuchar, Grassley, colleagues introduce bipartisan legislation to boost competition and rein in big tech', news release, Amy Klobuchar website, 15 June 2023

p. 181, the Digital Consumer Protection Commission Act: 'Warren, Graham unveil bipartisan bill to rein in big tech', press release, Elizabeth Warren website, 27 July 2023

p. 181, the highest in more than a decade: Dave Michaels and Ryan Tracy, 'Wall Street deal making faces greater scrutiny, delays under FTC's Lina Khan', *Wall Street Journal*, 15 August 2022

p. 182, When his appointment was announced: 'Trump names Andrew Ferguson as next chair of Federal Trade Commission', *Guardian*, 11 December 2024

p. 182, a Consumer Privacy Act: 'Understanding the California Consumer Privacy Act (CCPA)', Thomson Reuters, n.d.

p. 182, on First Amendment grounds in the Supreme Court: Bobby Allyn and Nina Totenberg, 'Supreme Court puts Florida and Texas social media laws on hold', NPR, 1 July 2024

p. 182, Arkansas: 'SB396 – To create the Social Media Safety Act . . .', Arkansas State Legislature, 94th General Assembly – Regular Session, 2023

p. 182, Utah: 'SB152 Social Media Regulation Amendments', Utah State Legislature, 2023 General Session

p. 182, Texas: 'Bill HB 18', Texas Legislature, 2024

p. 182, California: 'AB-2273 The California Age-Appropriate Design Code Act', California Legislative Information, November 2022

p. 182, Louisiana: 'HB61', Louisiana State Legislature, 2023 Regular Session

p. 182, And forty-one state attorneys-general got together: Cristiano Lima-Strong and Naomi Nix, '41 states sue Meta, claiming Instagram, Facebook are addictive, harm kids', *Washington Post*, 24 October 2023

p. 186, success stories like Volkswagen: 'Europe may fall behind in self-driving cars', *Automotive News Europe*, 6 August 2018

p. 186, it had dropped below 15 per cent: 'European Union: share in global gross domestic product based on purchasing-power-parity from 2019 to 2029' (chart), Statista, October 2024

p. 186, growing more slowly, generating poorer returns: Sven Smit et al., 'Securing Europe's competitiveness', McKinsey Global Institute, September 2022

p. 186, half of that in the States: 'GDP per capita, current prices' (map and list), International Monetary Fund, October 2024

p. 186, None of the top ten global companies: Daniel Liberto, 'Biggest companies in the world by market cap', Investopedia, updated 16 October 2024

p. 186, None of the dozen most valuable unicorns: 'The complete list of unicorn companies', CB Insights, December 2024

p. 186, Most of the world's AI foundation models: Nestor Maslej et al., 'The AI Index 2024 Annual Report', Institute for Human-Centered AI, Stanford University, April 2024

p. 187, Mario Draghi published a report: Mario Draghi, 'The future of European competitiveness', EU Commission, September 2024

p. 188, gave a speech at the Sorbonne: Gérard Araud, '"Our Europe is mortal. It can die." Decoding Macron's Sorbonne speech', *New Atlanticist*, 29 April 2024

p. 189, more than 900 million were eligible to vote: 'How strong is India's economy under Narendra Modi?', *The Economist*, 15 January 2024

p. 189, there were 692 million internet users in India: Simon Kemp, 'Digital 2023: India', Datareportal, 13 February 2023

p. 190, Indian tech companies have built billion-dollar companies: 'Unicorns in India: List of startup companies with unicorn status in 2024', *Forbes India*, 22 August 2024, updated 26 September 2024

p. 190, from 'welfare payments to loan applications': Benjamin Parkin et al., 'The India Stack: opening the digital marketplace to the masses', *Financial Times*, 20 April 2023

p. 190, with more than seventy unicorns: Leon Igel, 'Is India the world's new growth engine as many in the West believe?', NZZ, 8 August 2024

p. 190, a global powerhouse of AI: 'We are building next-gen AI to become a global powerhouse: Rajeev Chandrasekhar', *Economic Times*, 3 March 2023

p. 190, A large majority of Indians: Nestor Maslej et al., 'The AI Index 2023 Annual Report', Institute for Human-Centered AI, Stanford University, April 2023

p. 190, India has the second-largest developer base: Shilpa Phadnis, 'India users of GitHub expected to be at par with US number by 2025', *The Times of India*, 10 November 2022

p. 190, and third-largest start-up ecosystem: Indian government website, https://www.startupindia.gov.in/content/sih/en/international/go-to-market-guide/indian-startup-ecosystem.html

p. 190, could add nearly a trillion dollars: 'IndiaAI 2023', Ministry of Electronics and Information Technology, Government of India, October 2023

p. 191, When Google's Gemini AI model generated answers: Amrit Dhillon, 'India confronts Google over Gemini AI tool's "fascist Modi" responses', *Guardian*, 26 February 2024

8 The End of the Internet's Golden Age

p. 195, with officials reportedly stating: David Lawder, 'US drops digital trade demands at WTO to allow room for stronger tech regulation', Reuters, 26 October 2023

p. 198, According to a Reuters report: Ibid.

p. 199, the Freedom House think tank: Allie Funk and Jennifer Brody, 'Reversal of US trade policy threatens the free and open internet', Tech Policy Press, 15 November 2023

p. 199, Center for Strategic and International Studies: Erol Yayboke et al., 'The real national security concerns over data localization', CSIS, 23 July 2021

p. 205, Mark Montgomery and Theo Lebryk: Mark Montgomery and Theo Lebryk, 'China's dystopian "New IP" plan shows need for renewed US commitment to internet governance', Just Security, 13 April 2021

p. 206, According to the Internet Society: Dan York, 'What is a splinternet? And why you should be paying attention', Internet Society, 23 March 2022

9 Superpowered Superpowers

p. 211, vaulting chipmaker Nvidia to a $3.2 trillion valuation: Prarthana Prakash, 'Nvidia single-handedly beats Europe's biggest stock markets, Deutsche Bank finds, as London Stock Exchange chief says reform is on the way', *Fortune*, 21 June 2024

p. 212, The first BIS order implemented export controls: 'Commerce implements new export controls ...', press release, Bureau of Industry and Security, US Department of Congress, 7 October 2022

p. 213, as *Time* magazine put it: Gregory C. Allen, 'China is striking back in the tech war with the US', *Time*, 20 July 2023

p. 214, scholar Chris Miller: Chris Miller, 'What the most "Chinese" smartphone yet tells us about politics', *Financial Times*, 21 September 2023

p. 214, In 2023, the government proposed new rules: 'Measures on the administration of generative artificial intelligence services (draft for solicitation of comments)', China Law Translate, 11 April 2023

p. 214, Chinese LLMs, it is reported: Jared Cohen et al., op. cit.; see p. 153 above

p. 214, As *The Economist* explains: 'China is shoring up the great firewall for the AI age', *The Economist*, 26 December 2023

p. 215, US Treasury Secretary Janet Yellen warned: 'Remarks by Secretary of the Treasury Janet L. Yellen on the US–China economic relationship at Johns Hopkins School of Advanced International Studies', US Department of the Treasury website, 20 April 2023

p. 215, *The Economist* warned: 'The destructive new logic that threatens globalisation', op. cit.; see p. 177 above

p. 216, International trade law expert Anu Bradford: Anu Bradford, *Digital Empires: The Global Battle to Regulate Technology*, Oxford University Press, 2023

p. 217, Goldman Sachs's Jared Cohen and George Lee: Jared Cohen et al., op. cit.; see p. 153 above

p. 217, open-source LLMs, including Falcon and Jais: Simeon Kerr and Madhumita Murgia, 'UAE launches Arabic large language model in Gulf push into generative AI', *Financial Times*, 30 August 2023

p. 218, the margin of inequality between rich and poor nations: Cristian Alonso et al., 'How artificial intelligence could widen the gap between rich and poor nations', IMF Blog, 2 December 2020

p. 218, the Bank for International Settlements: Giulio Cornelli et al., 'Artificial intelligence, services globalisation and income inequality', BIS Working Papers No. 1135, Bank for International Settlements, October 2023

p. 218, MIT economist Daron Acemoğlu: Daron Acemoğlu, 'Remaking the post-Covid world', *Finance & Development*, vol. 58, no. 1, March 2021

p. 218, as Pew Research suggests: Janna Anderson et al., 'Experts say the "new normal" in 2025 will be far more tech-driven, presenting more big challenges', Pew Research Center, 18 February 2021

p. 219, the discovery of antibiotics: Gary Liu et al., op. cit.; see p. 119 above

p. 219, generate equally accurate and detailed MRIs: 'Accelerating MRI reconstruction via active acquisition', op. cit.; see p. 119 above

p. 219, help preserve the world's language diversity: 'Preserving the world's language diversity through AI', Meta, 22 May 2023

p. 221, As Meta's Chief Technology Officer ... explains: Reed Albergotti, 'Meta's CTO on how the generative AI craze has spurred the company to "change it up"', Semafor, 20 December 2023

p. 223, the Political Declaration on Responsible Military Use of Artificial Intelligence and Autonomy: US State Department website, https://www.state.gov/political-declaration-on-responsible-military-use-of-artificial-intelligence-and-autonomy-2/, 9 November 2023

10 *Let in the Sunlight*

p. 233, Benedict Evans describes: Benedict Evans, 'When tech says "no"', op. cit.; see p. 172 above

p. 234, report by the Tony Blair Institute for Global Change: Chris Yiu, 'A new deal for big tech: next-generation regulation fit for the internet age', Tony Blair Institute for Global Change, 1 November 2018

p. 236, Nate Persily told a Senate subcommittee: Jed Ngalande, 'Stanford law professor calls for digital platform transparency legislation in Senate testimony', *Stanford Daily*, 10 May 2022

p. 239, an idea first proposed by Mark Zuckerberg in 2018: Mark Zuckerberg, 'A blueprint for content governance and enforcement', Facebook post, 15 November 2018

p. 239, the Appeals Centre Europe: Cynthia Kroet, 'New appeals body to rule on Facebook, TikTok, YouTube's policy violation complaints', msn.com, 8 October 2024

p. 240, the ancient Athenian idea of sortition: Paul Cartledge, 'And the lot fell on . . . sortition in Ancient Greek democratic theory & practice', OUP blog, 31 March 2016

p. 240, a House of Commons committee inquiry: Stephen Elstub and Jayne Carrick, 'Evaluation of the citizens' assembly on the inquiry of long-term funding of adult social care', Newcastle University, February 2019

p. 241, a pilot about generative AI: Jennifer Broxmeyer, 'Leading the way in governance innovation with community forums on AI', Meta, 3 April 2024

p. 242, In her book of the same name: Helen Nissenbaum, *Privacy in Context: Technology, Policy, and the Integrity of Social Life*, Stanford University Press, 2009

p. 243, most audio data can be kept: 'Notice of monitoring and recording to improve safety in Horizon Worlds', Meta, n.d.

p. 245, the *MIT Technology Review*'s example: Anouk Ruhaak, 'How data trusts can protect privacy', *MIT Technology Review*, 24 February 2021

p. 245, the idea of 'middleware': Katharine Miller, 'Radical proposal: middleware could give consumers choices over what they see online', Institute for Human-Centered AI, Stanford University, 20 October 2021

p. 246, the Fediverse: Tom Coates, 'How Threads will integrate with the Fediverse', Plasticbag.org, 11 January 2024

p. 247, whether it should nationalise parts of the 5G network: Jonathan Swan et al., 'Scoop: Trump team considers nationalizing 5G network', Axios, 28 January 2018

p. 248, As the Carnegie Endowment for International Peace notes: Erik Brattberg et al., 'Europe and AI: leading, lagging behind, or carving its own way?', Carnegie Endowment for International Peace, 9 July 2020

p. 249, As Nate Persily remarked: 'Testimony of Professor Nathaniel Persily ... before the United States Senate Committee on the Judiciary, Subcommittee on Privacy, Technology, and the Law: "Platform transparency: understanding the impact of social media"', submitted 2 May 2022

11 A Deal to Save the Internet

p. 261, The OECD has forecast that such a tax: Emma Agyemang, 'OECD agrees global treaty targeting tax from digital giants', *Financial Times*, 11 October 2023

p. 261, President Trump withdrew US support for the deal: David Lawder, 'Trump effectively pulls US out of global corporate tax deal', Reuters, 21 January 2025

p. 263, the dollar's position as the global reserve currency: Will Rinehart, 'The specter of a trade war', Center for Technology, Science, and Energy, 10 February 2025

p. 263, Project 2025: 'Project 2025 takes on the World Bank and IMF – harbinger of an uncertain new era of geopolitics?', Bretton Woods Project, 16 October 2024

p. 269, the Indian government's interest in making AI chips: Alex Travelli, 'Modi wants to make India a chip-making superpower. Can he?', *New York Times*, 13 September 2023

Index